T3-BNZ-860

Solar Selective Surfaces

Alternate Energy
A WILEY SERIES

Series Editors:

MICHAEL E. McCORMICK

Department of Naval Systems Engineering
U.S. Naval Academy
Annapolis, Maryland

DAVID L. BOWLER

Department of Engineering
Swarthmore College
Swarthmore, Pennsylvania

Solar Selective Surfaces

O. P. Agnihotri and B. K. Gupta

Energy, The Biomass Options

Henry Bungay

Solar Selective Surfaces

O. P. AGNIHOTRI

Professor of Physics, Department of Physics

B. K. GUPTA

Senior Scientific Officer, Center of Energy Studies
Indian Institute of Technology
New Delhi, India

A WILEY-INTERSCIENCE PUBLICATION

JOHN WILEY & SONS **New York . Chichester . Brisbane . Toronto**

Copyright © 1981 by John Wiley & Sons, Inc.

All rights reserved. Published simultaneously in Canada.

Reproduction or translation of any part of this work beyond that permitted by Sections 107 or 108 of the 1976 United States Copyright Act without the permission of the copyright owner is unlawful. Requests for permission or further information should be addressed to the Permissions Department, John Wiley & Sons, Inc.

Library of Congress Cataloging in Publication Data:

Agnihotri, O P 1939-
Solar selective surfaces.

"A Wiley-Interscience publication."
Includes index.
1. Solar collectors. I. Gupta, B. K., 1954- joint author. II Title.

TJ810.A36 621.47 80-17392
ISBN 0-471-06035-6

Printed in the United States of America

10 9 8 7 6 5 4 3 2 1

621,47
A 223 5

To the memory of Roland S. Woolson
Originator of the Alternate Energy series

Dedicated to Professor M. S. Sodha

Series Preface

During the 1970s it became clear that the world's known nonrenewable energy resources are decreasing rapidly and may be exhausted within the foreseeable future. In response to this disturbing prospect, the technologically advanced countries began to focus attention on renewable resources. As a result, there have been significant advances in such areas as solar heating and cooling, photovoltaics, wind power, bioenergy, and ocean wave energy.

The purpose of Alternate Energy: A Wiley Series is to discuss solutions to the technological and economic problems associated with the widespread use of renewable energy resources. The series is intended to introduce readers to the range and potential of these resources, to describe currently available and anticipated methods for conversion and delivery, and to consider economic aspects. The authors published in this series are well known in their fields and have made significant contributions. We have planned the series to meet the needs of all those interested in alternate energy, both practitioners and students.

MICHAEL E. MCCORMICK
DAVID L. BOWLER

Annapolis, Maryland
Swarthmore, Pennsylvania

Preface

The world energy crisis has given a new impetus to the solar energy utilization research and development program all over the world. Terrestrial solar radiation is a low-intensity, variable energy source arriving at about 1000 W/m^2. The economic feasibility of solar energy utilization depends upon efficient collection, conversion, and storage. The efficient utilization of solar energy for heating, cooling, and process applications requires the use of flat-plate or focused collector systems which first capture as much as possible of incoming radiation and then deliver a high fraction of the captured energy to the user which is generally a working fluid. The conversion efficiency of a collector system is limited by the thermal losses from the heated absorber due to conduction, convection, and radiation. The losses become increasingly significant at higher temperatures. The economical and efficient utilization of thermal energy derived from solar radiation using solar collectors requires an efficient and low cost solar "selective coating" or "selective surface." An efficient solar selective surface is defined as having a high absorptance α over the solar spectrum (0.30–2.0 μm) and, in addition also having a low emittance ε to reduce thermal radiative heat losses. The achievement of such a surface with wavelength selective properties has been possible due to the fact that the solar spectrum and the thermal infrared spectrum of heated bodies do not overlap to any appreciable extent (for temperatures below 500°C, 98% of the thermal infrared radiation occurs at wavelengths greater than 2 μm). A parameter that has been used to characterize a solar selective surface is the ratio of solar absorptance to thermal emittance, α/ε. However, the trade-off between α and ε for various collector systems has been discussed by various authors and it has been shown that an increase of α is more effective in improving the operating efficiency than a corresponding decrease in ε.

Solar selective surfaces are generally classified into two categories: (1) the reflection–absorption type, i.e., surfaces which have high absorption in the solar spectrum and are good reflectors for heat waves, and (2) the reflection–transmission type, i.e., surfaces, which have high transmission in the solar spectrum but are good reflectors of thermal radiation. All the black absorbing type of coatings belong to category 1, and the transparent heat mirrors belong to category 2. While the absorbing types of surfaces are applied directly on the collector surfaces, the heat mirrors are applied to cover plates, which are separate from the absorber surface.

With the renewed interest in the solar energy program, there has been an upsurge of interest in the solar selective surfaces during the last few years. Although several review articles on black absorbing and transparent selective surfaces have appeared recently, there is at present no book which gives a coherent account of the preparation, properties, characterization, and application of these surfaces. Looking at the importance and the amount of interest in the subject, we therefore, saw a justification in undertaking such a venture.

The subject matter covered in this book can be divided into five chapters.

In the first chapter, we have discussed the preparation, properties, and the role of transparent conducting coatings as solar selective surfaces, with an emphasis on new work which has appeared in the literature after 1976. Complete review of the basic properties of the transparent conductors has not been given because of the existence of excellent reviews by Holland (25), Jarzebski and Marton (26), Haacke (24), and Vossen (23). Cadmium stannate is however, a recent development in the line of transparent conductors and holds promise for solar energy applications and has therefore been described in detail.

The second chapter describes the application of selective surfaces to photothermal conversion. No discussion on the selective surfaces would be considered as complete unless it describes the significant data available in the literature on the effect of a selective surface on the collection efficiency. Unfortunately, not enough data is available on the testing of selective surfaces under load conditions and more work is needed. However, the few results which have been published are adequately covered in this chapter.

The third chapter includes a discussion of transparent conductors used in photovoltaic energy conversion. Devices of the type ITO/Si, SnO_2/Si, ITO/GaAs, ITO/InP and SnO_2/GaAs have been the subject of recent investigation. Such devices are of current interest because they represent a potentially low cost method for fabricating large-scale solar energy conversion arrays. Oxide semiconductor solar cells have yielded efficiencies which are comparable to conventional diffused silicon solar cells. Some of the advantages of the transparent conducting oxide solar cells are: (1) the oxides exhibit indices of refraction in the right range to provide an inherent antireflection coatings on silicon; (2) the carrier generation occurs directly in the depletion region, increasing the quantum yield towards the short wavelength region; (3) the high optical transparency and low electrical resistivity greatly simplifies making the front contact and (4) the oxide absorbs in the ultraviolet and therefore acts as a window for sunlight. Moreover, in such cells, as the base doping increases, the dark current decreases, allowing the lifetime, collection depth, radiation resistance, and efficiency to increase.

The characterization of selective surfaces has been discussed in the fourth chapter. This chapter describes the various measurement techniques used for absorptance and emittance determination. Divergent values of absorptance and emittance have been reported in the literature depending upon the equipment and techniques used. Some investigators have used direct beam spectral reflectance measurements to characterize their surfaces, while others use hemispherical spectral reflectance and calorimetric methods. A critical evaluation of the various techniques was necessary and has been done in this book. Both direct and indirect determination methods have been discussed.

The fifth chapter describes in great detail the black solar selective absorbing surfaces. Both the intrinsic and composite materials and various techniques like electrodeposition, spray, sputtering, coversion coatings, absorber—reflector tandems by vacuum deposition, and semiconductor paints have been discussed in detail. Preparation of selective surfaces for solar concentrators by magnetron sputtering is a new development and is included in the book.

The sixth chapter gives the conclusions and recommendations.

The development of the subject matter of this book is partly based upon the work on selective surfaces and photovoltaics at the Center of Energy Studies and the Department of Physics at the Indian Institute of Technology, Delhi. In the preparation of this book, we have drawn heavily from a variety of sources including the developments of several groups and authors. They are: the work of H. Tabor and his colleagues, the work of G. Haacke and his colleagues, Gier and Dunkle's contribution, the publications of J.C.C. Fan and his colleagues, the experimental results on solar selective surface testing by R. M. Winegarner and his colleagues at the Optical Coatings Laboratory, J. B. DuBow and his group at Colorado State University, J. Shewchun and his group at McMaster University, Canada, J. A. Duffie and W. A. Beckman's group at the University of Wisconsin, the solar energy group of the Energy Research Center, University of Sydney, Glen McDonald's contribution from the NASA Lewis Research Center, B. O. Seraphin and A. B. Meinel's group at Optical Sciences Center, University of Arizona, R. L. Anderson and his group, M. Telke's contribution to selective surfaces and many other authors, the references of whose works have been cited.

This book presents an integrated picture of selective surfaces in which both the transparent and black surfaces and their applications have been discussed and is a first attempt of its kind. We have tried to include all the significant contributions in the field. We hope that the book will be useful to M.S./M.Tech./Ph.D students of solar energy and scientists and technologists engaged in the development of selective surfaces and photothermal applications of solar energy.

O. P. Agnihotri
B. K. Gupta

New Delhi, India
November 1980

Acknowledgments

The development of energy research programme at I.I.T. Delhi owes a great deal to the leadership and vision of Prof. M. S. Sodha, the founder head of the Center of Energy Studies and the Deputy Director of Indian Institute of Technology (I.I.T.), Delhi. The nucleus for energy research was created by him when he collected enthusiastic faculty from the departments of Sciences and Engineering who were interested and already contributing in the areas of Solar Energy, MHD Power Generation, Laser Plasma Interaction, Fuel and Coal, Power Systems, and Refrigeration and Air-conditioning to form an interdisciplinary group. It was mainly through his efforts that this activity was recognized by the Institute and the Government first as a School and then as a Center of Energy Studies. It was under his inspiration that we undertook the work on Solar Selective Surfaces and Oxide Semiconductor Solar Cells under the Energy Research Programme. The suggestion to write this book came from him and we are grateful to him for the support and encouragement and his deep interest in the progress of this book. The foundation of the solar energy work at I.I.T. Delhi was laid by Prof. S. C. Jain, former Head of the Physics Department and the Dean of Science and the present Director of the Solid State Physics Lab. Delhi, when the first silicon solar cell was fabricated in 1966 at the Institute under his leadership. We are extremely grateful to him for his kind interest, constant encouragement and valuable guidance.

We acknowledge the support and encouragement given by Prof. S. S. Mathur, Head of the Center of Energy Studies and the Co-ordinator of Solar Energy Programme throughout the progress of this work. We are grateful to many colleagues and friends in particular to Prof. P. K. C. Pillai, Prof. B. B. Tripathi, Prof. A. B. Bhattacharyya and Prof. H. P. Garg who gave all possible help and encouragement. We acknowledge the constructive criticism and contributions of graduate students in the Physics Department and Center of Energy Studies. The encouragement and support of Prof. O. P. Jain, the Director of the Institute is gratefully acknowledged.

We appreciate the co-operation of many authors and publishers who have readily given us permission to use drawings or other material from their work. We have attempted to reference adequately all such sources.

We are also grateful to our family members for their patience while we were busy writing this book.

Finally, we acknowledge the assistance of Mr. S. D. Malik in the efficient typing of this book and Mr. N. S. Gupta for preparing excellent drawings on very short notice.

O. P. A.
B. K. G.

Contents

Solar Selective Surfaces

1

Transparent Conducting Coatings

Widespread utilization of solar energy requires the development of durable, low cost, optically efficient solar selective coatings. Solar selective surfaces in addition to having high absorptance and low thermal emittance, must be stable at high operating temperatures and resistant to atmospheric corrosion. At low operating temperatures ($< 60°C$) nonselective absorbers can serve the purpose, but for driving heat pumps efficiently and for cooling, higher operating temperatures are required. The use of optically efficient solar selective coatings is then necessary. Two approaches for solving this problem have been used. The first is to use an absorbing coating on the surface of the collector and the second is to use a transparent conducting coating on the cover of the solar collector. There are several advantages in following the second approach. The transparent conducting coating applied to the inner glass cover is not exposed to the same humidity and temperature conditions as the selective absorber coatings on the absorber surface. Heat mirrors have therefore shown better resistance to humidity, temperature, and durability as compared to selective absorbers.

For transparent conducting coatings, one can either use a metal with high infrared reflectivity and very low visible absorption or a semiconductor with a band gap sufficiently wide to be transparent in the visible range and a high enough carrier concentration to have high infrared reflectively. A brief review of the work on metal films and wide band gap semiconductor coatings having desirable optical properties for applications in solar collector systems has been given below. The economic feasibility for the use of transparent conducting coatings requires low-cost, large-area techniques. Recent work on the oxide–semiconductor coating techniques which will have important bearing on solar energy applications are also described. A selective absorber surface is characterized by two parameters, namely, absorptivity and emissivity. The equivalent parameters for a transparent conducting coating are the effective absorptivity and the effective emissivity. These values have been calculated for various transparent conductors and the data have been given to show the effectiveness of the heat mirrors compared to selective absorbers.

1.1 Optical Coatings for the Collection and Conservation of Solar Energy

Recent experiments (1–12) have established that transparent heat mirror coatings that transmit in the visible and reflect in the infrared will have important applications in the collection and conservation of solar energy. The predominant heat-loss mechanism in a flat plate collector is the radiation from the absorber surface. The efficiency of a collector system for thermal conversion of solar radiation can be increased by increasing the operating temperature of the collector. This can be done by reducing the infrared emission from the collector. Solar energy collectors with a black absorbing coating have been developed which have high absorptivity in the visible but have low infrared emissivity. The coatings in this case will however be exposed to the same temperature cycle to which the collector surface is exposed. The collector must be stable at the stagnation temperature and should be able to withstand the excursions from daytime temperature to winter night time temperature. In contrast to the selective absorber coating which is applied to the absorber plate, the transparent heat mirror coating is applied to the glass cover of the flat-plate collector. The heat mirror permits the transmission of solar radiation but reflects the infrared radiation. For zero transmission, the heat loss by radiation is proportional to $1 - R$, where R is the infrared reflectivity. The transparent conductor should have reflectivity as close to 100% as possible.

The transparent heat mirror should transmit solar radiation (with a wavelength range of $0.4 < \lambda < 2.5\mu m$, where λ is the wavelength) but reflect the thermal radiation from the heated absorber ($2.5 < \lambda < 100\mu m$). Two types of materials have been used. One type is a semiconductor with a sufficiently wide band gap to be transparent in the visible and high enough carrier concentration to have high infrared reflectivity. The second type of material used is a metal with high infrared reflectivity and low enough visible absorption for thin films to be transparent. The visible absorption of metal films is limited by their reflectivity. Therefore their effectiveness as transparent heat mirrors can be improved if they can be coated in a manner that reduces their reflectivity in the visible without affecting their infrared reflectivity.

1.2 Metal Films with High Infrared Reflectivity

Thin metal films that have a high reflectivity for IR radiation provide important thermal insulation. Although several metal films have been tried, acceptable transparent electrode properties have only been achieved in gold films. Thin films of silver and copper show aging effects when exposed to atmosphere (13, 14). Gold films have been used as thermal shields in the steel industry (15) and as insulation in transparent furnaces (16). The thickness of

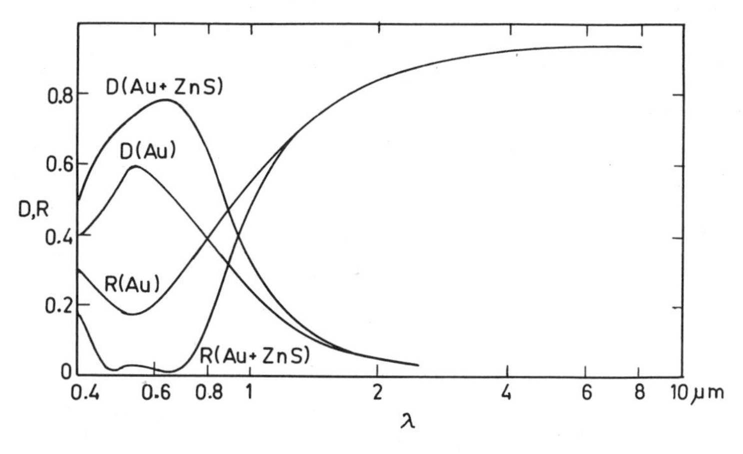

Fig. 1.2.1 Spectral transmission D and reflection R of a 13 nm thick gold film. Au: gold film alone. Au + ZnS: gold film coated with antireflection layer of ZnS. Adapted from Groth and Kauer (1965) (Ref. 17).

the gold layer has to be optimized because if the layer is too thick, it will transmit only little light while if too thin will not reflect well in the infrared. The optimum thickness of gold film is around 15 nm (17) as the reflection coefficient R approaches 90% even for the near infrared while the transmission for the sodium D lines remains larger than 50%. The visible reflection at the metal surface can be greatly reduced and thus transmission enhanced by coating the metal film with a dielectric layer (18). The effect of coating a 32 nm ZnS film on to a 13 nm gold film on the visible transmission and reflectivity is shown in Fig. 1.2.1 (17). The dielectric layer increases the transmission from 57 to 77% without reducing the reflection coefficient in the infrared. The refractive index of the dielectric chosen has to be higher than the metal film. Holland and Siddal (19) have reported that the visible transmission through thin Au films can be greatly enhanced by sandwiching the Au between two antireflection layers of Bi_2O_3. This follows an earlier observation of Gillham and Preston (20, 21) that thin Au films were much more conductive when a thin Bi_2O_3 layer is first sputtered on to a glass plate. The sputtered Bi_2O_3 was found to be amorphous thus containing large number of point defects than a glass surface. This leads to an early coalescence into a continuous film (22).

Fan et al. (1) have fabricated transparent heat mirrors of $TiO_2/Ag/TiO_2$ on glass with a visible transmission of 84% (at 0.5 μm), [the peak of the emission curve (23) for a room temperature blackbody] by RF sputtering. Silver films have lower visible absorption than gold films of the same thickness. The measured reflectivity and transmission of a 180 Å TiO_2/180 Å Ag/180 Å TiO_2 film on glass substrate is shown in Fig. 1.2.2 (1). The films have excellent optical properties and are adherent and stable up to 200°C and offer great promise for applications in solar thermal power conversion

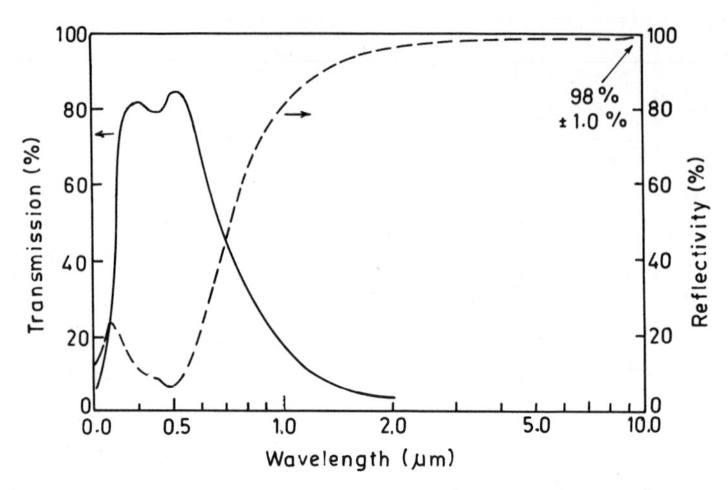

Fig. 1.2.2 Measured reflectivity and transmission of 180 Å TiO_2/180Å Ag/180Å TiO_2 film on a CG 7059 glass substrate ($\approx$ 1 mm thick). Adapted from Fan et al. (1974) (Ref. 1).

and as transparent thermal insulators for the windows of residential and commercial buildings.

1.3 Wide Band Gap Semiconductors as Heat Mirrors

Wide band gap semiconductors, which will transmit the visible but will have high infrared reflectivity when suitably doped have been successfully used as heat mirrors. Free charge carriers in a semiconductor have a marked influence on the optical properties of a medium. Suppose that a medium has a dielectric constant ε_g in the absence of free charge carriers. If we bring in free charge carriers in a material by suitable doping then after Drude, the following dispersion relations will apply;

$$n^2 - k^2 = \varepsilon_g - \frac{Ne^2}{\varepsilon_0 m^*(\omega^2 + \gamma^2)} \tag{1.3.1}$$

$$2nk = \frac{\gamma N e^2}{\varepsilon_0 m^* \omega(\omega^2 + \gamma^2)} \tag{1.3.2}$$

The value of damping factor γ depends upon the D.C. mobility μ of charge carriers. $\gamma = e/\mu m^*$. The quantity N is the concentration of free charge carriers, e is the charge on the electron, m^* is the effective mass of the carriers and ε_0 is the dielectric constant of free space. The two expressions may be rewritten introducing the plasma frequency:

$$\omega_p = \left(\frac{Ne^2}{\varepsilon_0 \varepsilon_g m^*} - \gamma^2\right)^{1/2} \tag{1.3.3}$$

This is defined by putting $n^2 - k^2 = 0$ in the first of the above dispersion relations. The formulae are then simplified (17) with ω/ω_p as the independent variable and γ/ω_p as the material parameter:

$$n^2 - k^2 = \varepsilon_g\left[1 - \frac{1 + (\gamma/\omega_p)^2}{(\omega/\omega_p)^2 + (\gamma/\omega_p)^2}\right] \tag{1.3.4}$$

$$2nk = \frac{\varepsilon_g(\gamma/\omega_p)[1 + (\gamma/\omega_p)^2]}{(\omega/\omega_p)[(\omega/\omega_p)^2 + (\gamma/\omega_p)^2]} \tag{1.3.5}$$

From these expressions n and k can be calculated and then the reflection coefficient R can be evaluated from

$$R = \left[\frac{(n-1)^2 + k^2}{(n+1)^2 + k^2}\right] \tag{1.3.6}$$

The results of such a calculation are shown in the Fig. 1.3.1. The steepness of cutoff depends upon the ratio γ/ω_p. To obtain a filter with sharp cutoff, the ratio γ/ω_p should be as small as possible. The position of cutoff wavelength is primarily determined by ω_p, provided the film thickness is correctly chosen. In order to make γ/ω_p small, γ itself must be small. This means that we require a medium in which the free charge carriers have the largest possible product of mobility and effective mass. In addition we require a high concentration of free charge carriers, N, which should be at least 10^{20} cm^{-3}. It is not so easy to achieve this concentration because most semiconductors

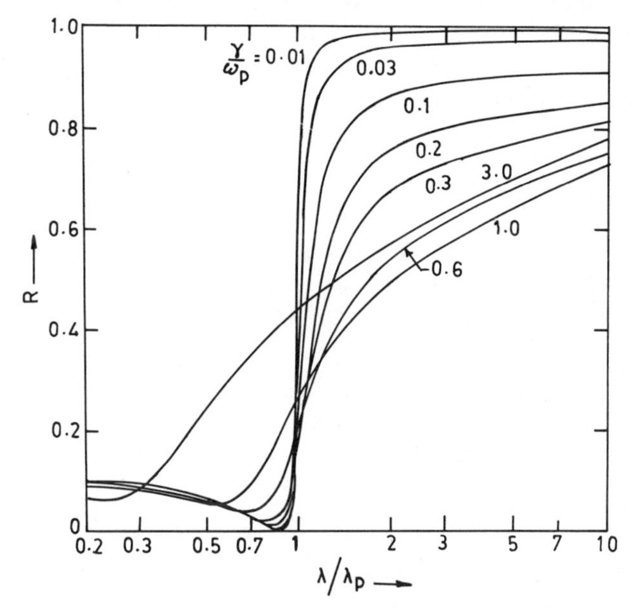

Fig. 1.3.1 Reflection coefficient for a medium containing free charge carriers. The material is defined by the relative dielectric constant of the medium $\varepsilon_g = 4$, and by the plasma wavelength $\lambda_p = 2\pi c/\omega_p$ where ω_p is the plasma frequency. Adapted from Groth and Kauer (1965) (Ref. 17).

cannot be doped so strongly. However, in some semiconductor oxides like SnO_2 and In_2O_3, a concentration higher than 10^{20} cm^{-3} can be achieved. In_2O_3 has a higher value of mobility for the free carriers as compared to SnO_2 and thus shows a promise of better results than those obtained with SnO_2. Yet another transparent conductor, Cd_2SnO_4 has received considerable attention by Haacke and his coworkers (10, 12, 24) who achieved in this material, transparency and conductivity better than In_2O_3 and have demonstrated the potential of Cd_2SnO_4 as a low cost material for solar energy applications.

The published work on transparent conductors up to 1955 has been reviewed by Holland (25). Because of numerous applications to which these materials are put, a substantial amount of new work has appeared since then. Jarzebski and Marton (26) have brought out an extensive review article summarizing the structural, electrical, and optical properties of SnO_2. Evaluation of transparent metal and semiconductor coatings including Cd_2SnO_4 has been discussed by Haacke (24) in an excellent review article. A detailed review article by Vossen (23) has recently appeared which describes the present status and technical applications of "Transparent Conducting Films." Metal films, such as Au, Ag, and Pt, and oxide films based upon, for example, SnO_2, In_2O_3, and CdO are considered. The article has discussed the preparative methods employed, the physics of conductivity and transparency, dependence of electrical and optical properties on conditions of deposition, and areas of application. The review articles adequately cover the status of the transparent conductors till 1976–1977. In the present discussion therefore, we will confine ourselves to some of the properties relevant to solar energy applications and the new work which has appeared after 1976–1977.

1.4 Tin Oxide

SnO_2 has a tetragonal rutile structure having a space group $D_{4h}^{14}[P4_2/mmm]$ (27). The unit cell of the crystal structure of SnO_2 is shown in Fig. 1.4.1. The lattice parameters of SnO_2 are $a = b = 4.737$Å and $c = 3.185$ Å (28). The c/a

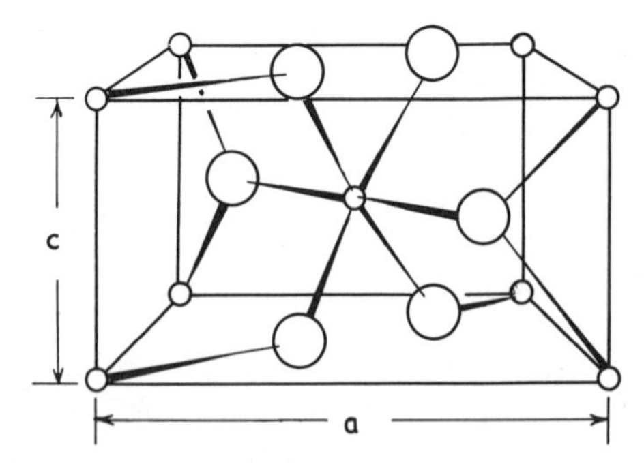

Fig. 1.4.1 Unit cell of the crystal structure of SnO_2. The large circles indicate oxygen atoms and the small circles indicate tin atoms.

ratio is 0.673. The ionic radii of O^{2-} and Sn^{4+} are 1.40 and 0.71 Å, respectively (29). It is *n*-type wide gap semiconductor. The native defects in pure crystals have been identified as doubly ionized oxygen vacancies with ionization energies of 30 and 150 MeV (30). Antimony and chlorine are convenient donors and provide free electrons (31, 32). Direct energy gap values of 3.57 and 3.93 eV have been found for light polarized perpendicular and parallel to the *c* axis, respectively (33), in SnO_2 crystals. These band gap values allow SnO_2 to be transparent in the entire visible region of the spectrum. The electrical and optical properties of doped SnO_2 films have been investigated by various authors (31, 32, 34–38). The absorption edge of thin SnO_2 films is around 3.7 eV, the position depending upon the free electron concentration. The mobility is around 10 cm^2/V-sec at 300°K and the highest electrical conductivities are in the 1200–1400 ohm^{-1} cm^{-1} range (24, 34, 39). The presence of secondary phases like SnO, Sn_2O_3 (40), and Sn_2O_4 in antimony doped films (41) has been reported and adversely affect the transparent conducting property of SnO_2 phase. Jordan (42) achieved a sheet resistance of 12 ohm/sq. with SnO_2 film deposited on soda lime glass by spray deposition. [The sheet resistance is given in ohm per square (ohm/sq.) to indicate that it measures resistance of a film with square surface area.] Although a variety of methods have been used for SnO_2 deposition, chemical vapor deposition (CVD) and reactive sputtering are the most promising ones.

1.5 Indium Oxide

The crystal structure of In_2O_3 is *bcc* with the lattice parameter value of 10.118 Å (27). It is a degenerate extrinsic *n*-type semiconductor. From the optical measurements on In_2O_3 crystals, Weiher and Ley (43) identified the onset of direct transitions at 3.75 eV. They also observed an indirect forbidden transition with an energy gap of 2.619 eV. Muller (44) reported an energy gap of 3.65 eV from measurements on thin films. Raza et al. (45) interpreted their optical absorption data on thin In_2O_3 layers as due to direct allowed (3.55 eV) and indirect forbidden (2.4 eV) transitions. Groth (46) reported an electron mobility of 75 cm^2/V-sec in tin doped In_2O_3 at the electron concentration of 10^{20} cm^{-3}. The titanium and zirconium dopings were found to give even higher mobilities of 120 and 170 cm^2/V-sec, respectively. The films in these experiments (46) were prepared by spray pyrolysis. The transparent conductor properties in tin doped In_2O_3 films were found to be the best. With tin, a free electron density of about 10^{21} cm^{-3} can be achieved which corresponds to a plasma oscillation with a characteristic wavelength in the near infrared. Kostlin, et al. (47) investigated optical and electrical properties of tin doped In_2O_3 films prepared by spray deposition. The films were electrically conductive and transparent in the visible range of the spectrum. The free electron concentration was observed to increase with tin content upto a maximum. The films act as an optical filter which transmits visible and reflects infrared radiation. Such a filter is used in low-pressure sodium lamps in which it reduces the energy losses due to radiation and thus increases the efficiency of lamps (48). The wavelength selectivity exhibited by tin doped In_2O_3 films makes them very suitable as selective surfaces which transmit the solar region of the spectrum but reflect the infrared radiations. The spectral reflection and transmission of two In_2O_3:Sn films of identical thickness (0.3 μm) but different free electron density produced by spray pyrolysis (47) are shown in the Fig. 1.5.1. The plasma reflection edge shifts with carrier concentration and the properties of tin doped In_2O_3 can therefore be made to correspond to an ideal heat mirror (Fig. 1.5.2).

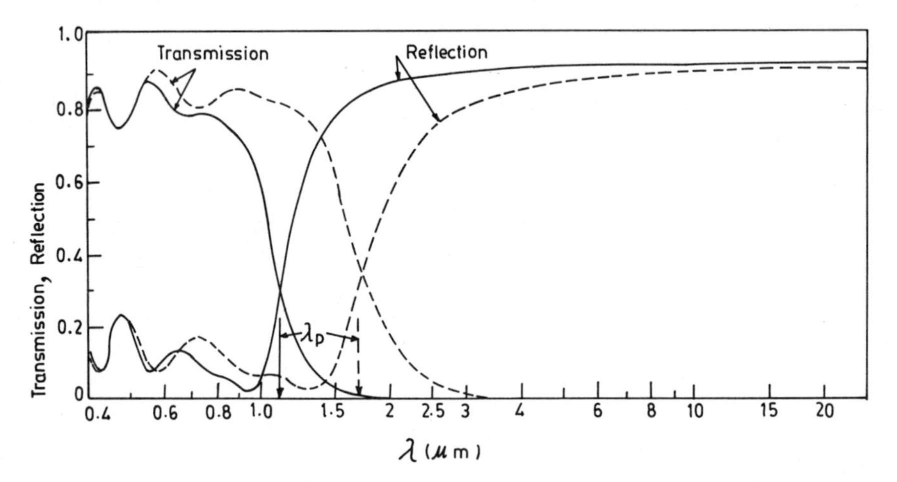

Fig. 1.5.1 The spectral reflection and transmission of two In_2O_3:Sn films of identical thickness (0.3 μm) but different free electron density; dashed curve: tin concentration 2%, $N = 0.3 \times 10^{21}$ cm^{-3}; drawn curve: tin concentration 8%, $N = 1.3 \times 10^{21}$ cm^{-3}. Adapted from Kostlin et al. (1975) (Ref. 47).

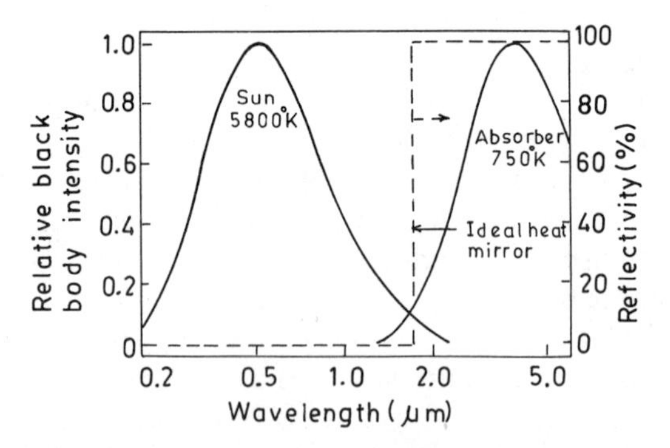

Fig. 1.5.2 Normalized distribution of radiant energy for blackbody temperatures of 5800°K (solar distribution at AMO) and 750°K. The dashed line shows the reflectivity of an ideal heat mirror.

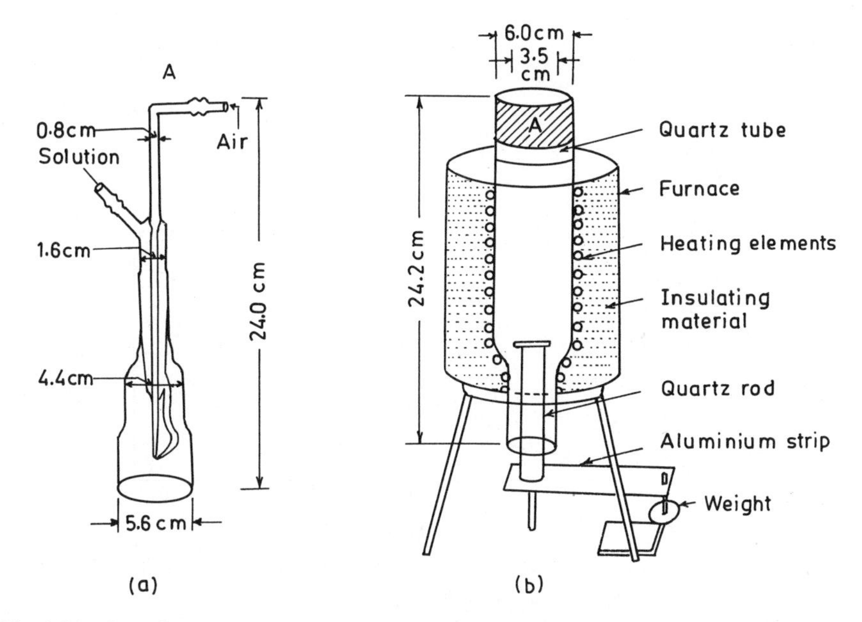

Fig. 1.5.3 Setup for spray deposition of In_2O_3 films. Adapted from Raza et al. (1977) (Ref. 45).

A process for producing In_2O_3 layers starting from indium has recently been described by Raza et al. (45). The thin-layer samples were prepared by a pyrolytic method in which an aerosol stream containing an atomic solution of $InCl_3$ in butyl acetate (293.3 g of $InCl_3 \cdot 4H_2O$ dissolved in 1 liter of butyl acetate) is blown on to a heated glass substrate. A suitable double-nozzle spray having a spray rate of 30 ml min^{-1} was used. In an oxidizing atmosphere at about 500°C, In_2O_3 films grow which adhere strongly to the glass substrate. The spray setup is shown in Fig. 1.5.3. Figure 1.5.3(b) gives the design of a special furnace used for spray deposition. A transparent quartz muffle 24.2 cm in length and 6 cm in diameter at the top and 3.5 cm at the bottom was used. The furnace is capable of giving 600°C at the position of the slide which is placed on the top of the quartz push rod. The quartz push rod can be placed in position by means of a weight and an aluminum strip. Figure 1.5.3(a) gives the diagram of a sprayer which fits on to the top of the muffle at A. Controlled additions of $SnCl_4 \cdot 5H_2O$ were made to spray solution to affect a change in the tin content of the films (49). The effects of adding up to 50 mole % of tin to pyrolytic In_2O_3 selective coatings were investigated. Up to 10% doping the resistivity is found to decrease with the corresponding increase of carrier concentration. Further additions increase the resistivity and decrease the carrier concentration (Fig. 1.5.4). The tin additions cause the lattice parameter to decrease, indicating that the decrease in resistivity is caused due to the replacement of In^{3+} by Sn^{4+} ions. The increase of resistivity at higher tin additions is due to poor crystallanity.

A number of investigators have discussed the properties of these films

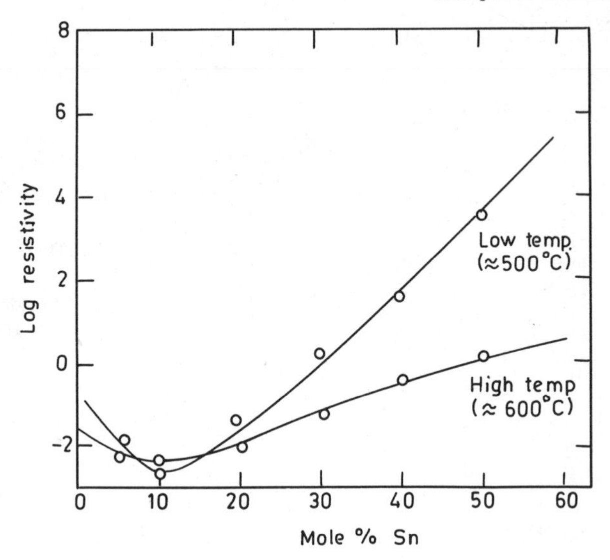

Fig. 1.5.4 Log resistivity as a function of mole % of Sn. The substrate temperatures are indicated. Adapted from Agnihotri et al. (1978) (Ref. 49).

prepared by sputter deposition (50–54). Fraser and Cook (52) reported the DC sputtering of $In_2O_3 + 9$ mole % SnO_2 in pure argon. They found that increasing the substrate temperature during film deposition lowered the resistivity. The highest reported electrical conductivity is 5600 ohm^{-1} cm^{-1} for a substrate temperature of 500°C. Films with sheet resistance of 2–3 ohm/sq. yielded average transmission values of about 80%. The films were found to be good infrared reflectors (≈ 90% reflectance) for wavelengths from 2.5 to 15 μm. Reflection characteristics of DC sputtered In_2O_3:Sn films are shown in Fig. 1.5.5.

Fan and Bachner (50) described a RF sputtering process for preparing tin

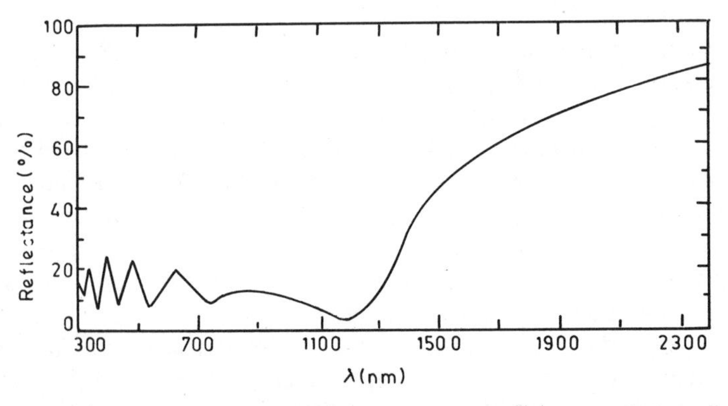

Fig. 1.5.5 Reflection characteristics of D.C. sputtered In_2O_3:Sn films. Adapted from Fraser and Cook (1972) (Ref. 52).

doped In_2O_3 films. Electrical resistivity of 2×10^{-4} ohm cm and 93% reflectivity at 10 μm has been reported. Electron mobilities of approximately 40 cm^2/V-sec at 7×10^{20} cm^{-3} free carriers have been observed. An index of refraction of 2.0 at 5500 Å has been reported. Specific heat treatment increases the index of refraction to 2.5 (54).

Solar energy conversion applications require film deposition over large area substrates, and in such cases CVD techniques can be useful. However for very good transparency, high infrared reflectivity, and low sheet resistance, sputtering techniques should be used.

1.6 Cadmium Stannate

Cd_2SnO_4 is an *n*-type defect semiconductor in which oxygen vacancies are believed to provide the donor states. The oxygen vacancy concentration can be varied over a wide range, resulting in a correspondingly wide range of conductivities. A large Burstein shift (55) has also been observed in the optical spectra, which indicates a low free carrier effective mass. Cd_2SnO_4 crystallizes in the orthorhombic structure (56, 57).

The utility of Cd_2SnO_4 for transparent electrodes was first indicated by Nozik (58) who measured electrical and optical properties of powder samples and thin amorphous films prepared by RF sputtering. Nearly stoichiometric films have a low band gap of 2.06 eV leading to an absorption edge in the middle of the visible spectrum. When films were deposited in an argon-rich $Ar–O_2$ mixture, a large Burstein shift of the absorption edge to about 2.9 eV

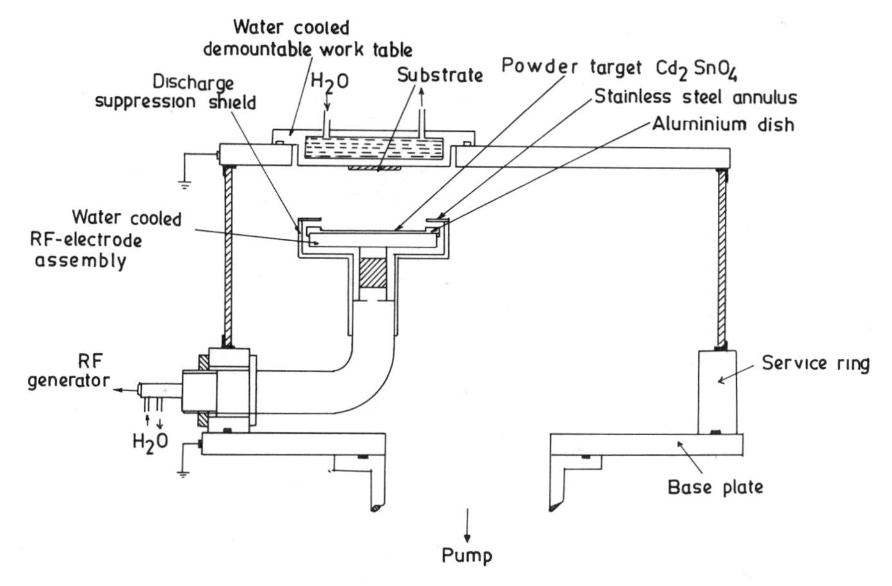

Fig. 1.6.1 Sputtering chamber. Adapted from Lloyd (1977) (Ref. 60).

was observed. An effective electron mass value of $0.04\ m_0$ is suggested, which is rather a low for a wide band gap oxide semiconductor.

Notable contributions on Cd_2SnO_4 come from Haacke et al. (59) and Lloyd (60) who studied optical and electrical properties of polycrystalline Cd_2SnO_4 films prepared by RF sputtering. Figure 1.6.1 shows the sputtering chamber and electrode system used by Lloyd (60). RF power at 13.56 MHz was supplied from a crystal controlled generator via a matching unit (R.D. Mathis Co., SG-125 and M 750). The RF electrode was a CVC Scientific Products Ltd., Part No. 276071. The substrate to target distance was 4 cm. Prior to film deposition, the chamber was evacuated to a pressure of 2×10^{-7} torr. Sputtering was carried out in an 80:20 Ar:O atmosphere at a pressure of 1.2×10^{-2} torr, at an RF power input of between 50 and 800 W. For a given target electrode and substrate geometry, the RF power input can be adjusted to a particular level to obtain stoichiometric films with minimum resistivity of 3×10^{-4} ohm-cm. Sputtered crystalline Cd_2SnO_4 films were found to contain CdO and $CdSnO_3$ phases which can be avoided by adjusting the deposition conditions. Cd_2SnO_4 with excellent transparency and sheet resistivity have been produced by Haacke et al. (59) and Lloyd (60) by RF sputtering from Cd_2SnO_4 powder. A sheet resistance of 1 ohm/sq. and 85% average transmission at 5500 Å has been achieved. Optical transmission of such films on silica substrates is shown in Fig. 1.6.2. Figure 1.6.3 shows the influence of substrate material on the optical transmission of Cd_2SnO_4. The transmission of films deposited on sapphire substrates is remarkable.

Haacke (10) has evaluated RF sputtered Cd_2SnO_4 films for applications in solar heat collectors. It has been demonstrated that these coatings are

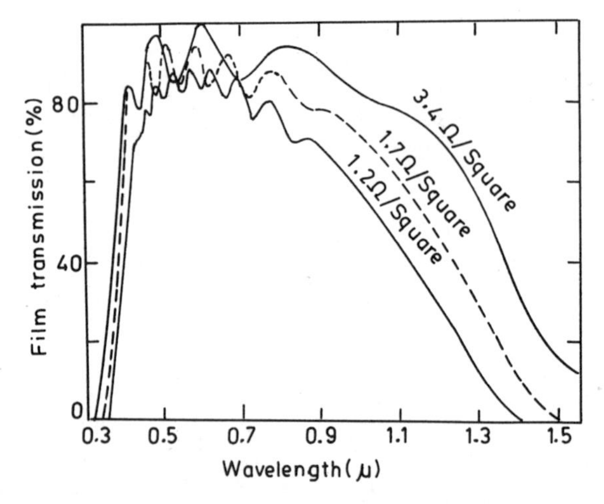

Fig. 1.6.2 Optical transmission versus wavelength of three cadmium stannate films with different electrical sheet resistances. The spectra were measured in a Cary 14 spectrophotometer with a blank substrate positioned in the reference beam. Adapted from Haacke (1976) (Ref. 9).

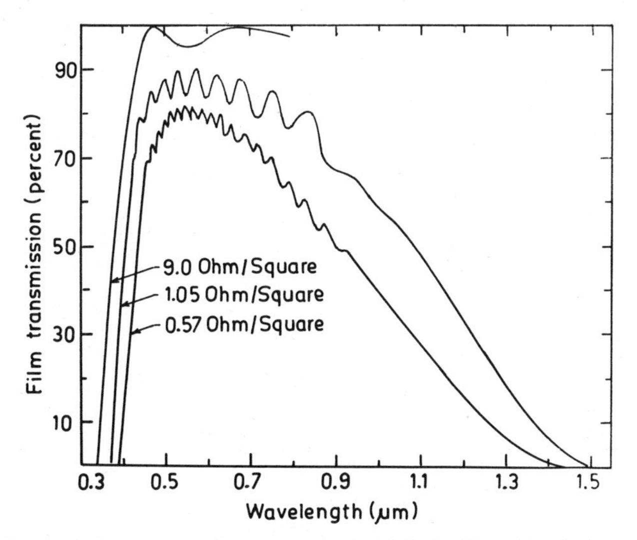

Fig. 1.6.3 Transmission spectra of sputter coated Cd_2SnO_4 films deposited on to different substrate materials; 9.0 ohm/sq. film on single crystal sapphire, 1.05 ohm/sq. film on silica, 0.57 ohm/sq. film on Corning 1720. Adapted from Haacke (1977) (Ref. 24).

potential candidates for transparent heat reflectors and selective absorber components in solar heat collectors. The total infrared emissivity of these coatings is of the order of 0.1 and the solar transmission is of the order of 0.78. By adjusting the post deposition annealing temperature solar absorptivity of ≈ 0.9 and total infrared emissivity of ≈ 0.1 can be obtained in Cd_2SnO_4/Si solar selective absorbers.

Agnihotri et al. (61) prepared cadmium stannate selective coatings by a pyrolytic method in which two parts of 0.2 *M* $CdCl_2 \cdot 5H_2O$ solution in ethyl alcohol and one part 0.2 *M* solution of $SnCl_4 \cdot 3H_2O$ in acetic acid has been sprayed on to a heated glass or quartz. The films were transparent in the visible portion of the spectrum and exhibited interference colors. The absorption data is interpreted as being due to an allowed direct transition of energy gap of 2.8 eV and an indirect interband transition of energy gap of 2.2 eV. The absorption coefficient reaches values exceeding 10^5 cm^{-1} at the higher photon energies, which is expected for direct allowed transitions. Percentage transmission of Cd_2SnO_4 films is shown in Fig. 1.6.4. Haacke et al. (12) prepared transparent Cd_2SnO_4 films by spray deposition. By adjusting the deposition conditions, both Cd_2SnO_4 and Cd SnO_3 can be deposited. The spray set used by them is shown in Fig. 1.6.5. The spray nozzle is located above a heated substrate which rests on a carbon plate. The plate is supported by fire bricks and is heated by a propane burner. The Cd_2SnO_4 phase is formed at substrate temperatures higher than 800°C while at lower temperatures (< 700°C), the $CdSnO_3$ phase is present. The film had a sheet resistance of 100 ohms/sq. but did not contain CdO.

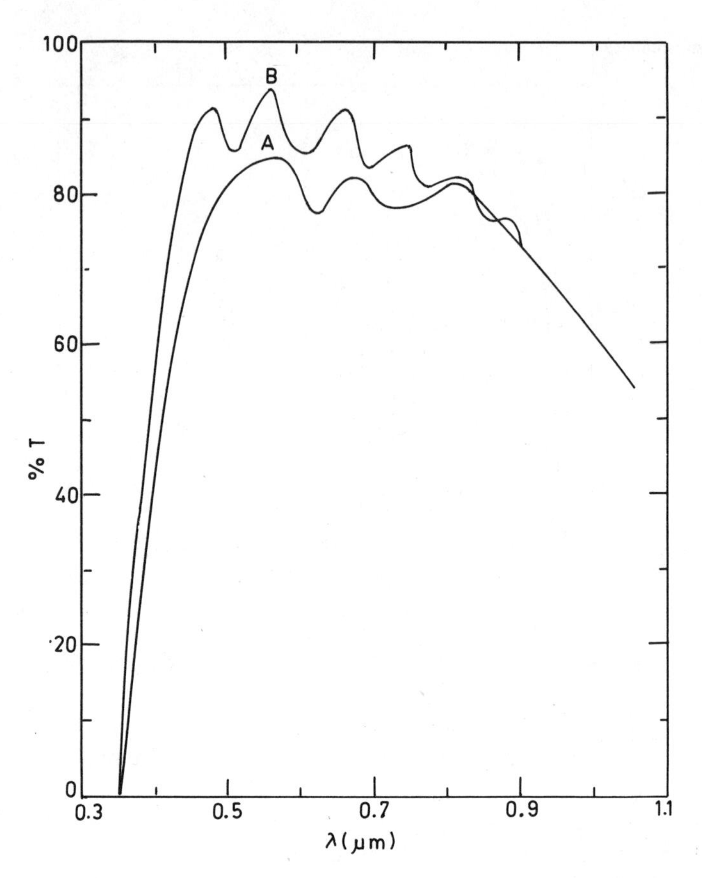

Fig. 1.6.4 Optical transmission versus wavelength of Cd_2SnO_4 films. Curve B is taken from Haacke (1976) (Ref. 9) and curve A is from Agnihotri et al. (1978) (Ref. 61).

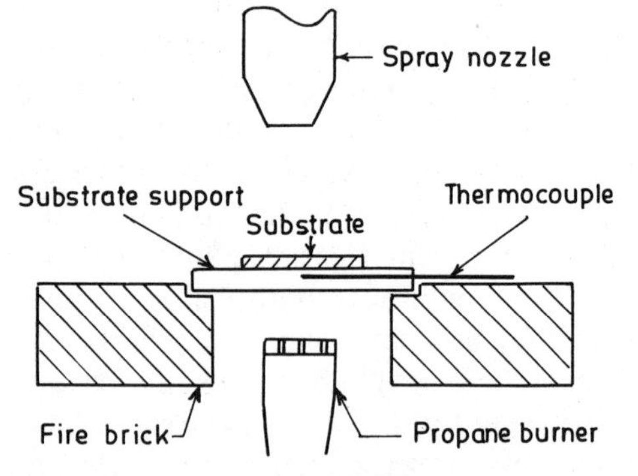

Fig. 1.6.5 Principal arrangement of spray equipment. Adapted from Haacke et al. (1977) (Ref. 12).

The adherence of these films on glass is good and they do not deteriorate with time and under humidity.

1.7 Evaluation of Transparent Conductors

A transparent conductor is evaluated from its optical transmission and conductivity both of which should be as large as possible. Haacke (11) defines a term ϕ_{TC} called the figure of merit which is used to evaluate a transparent conductor.

$$\phi_{\mathrm{TC}} = \frac{T^{10}}{R_s} \tag{1.7.1}$$

where T is the optical transmission and R_s is the electrical sheet resistance. If reflection losses from the transparent surface are negligible, we can write T as.

$$T = \exp(-\alpha t) \tag{1.7.2}$$

and

$$R_s = \frac{1}{\sigma t} \tag{1.7.3}$$

where α is the optical absorption coefficient in cm^{-1}, and t is the film thickness in cm. The dimension of the sheet resistance is given in ohm/sq. Combining eqs. 1.7.1, 1.7.2, and 1.7.3 we have

$$\phi_{\mathrm{TC}} = \sigma t \exp(-10\,\alpha t) \tag{1.7.4}$$

The above equation correlates the figure of merit ϕ_{TC} with material parameters α and σ. If reflection losses cannot be neglected, eq. 1.7.2 becomes (62)

$$T = (1 - R^2)[\exp(\alpha t) - R^2 \exp(-\alpha t)]^{-1} \tag{1.7.5}$$

where R is the reflectivity, ϕ_{TC} then becomes

$$\phi_{\mathrm{TC}} = \sigma t\{(1 - R^2)[\exp(\alpha t) - R^2 \exp(-\alpha t)]^{-1}\}^{10} \tag{1.7.6}$$

Equation 1.7.6 can be used to predict the transparent electrode properties of a material from the fundamental material parameters, α, σ, and n.

1.8 Transparent Heat Mirrors and Selective Absorbers

Solar selective absorbers are compared and evaluated by the ratio α/ε where α is the solar absorptance and ε is the total hemispherical emittance of the absorber surface. Fan and Bachner (2) have defined two heat mirror parameters, the effective solar absorptivity (α_{eff}) and the effective IR emissivity ($\varepsilon_{\mathrm{eff}}$), that are analogus to α and ε for selective absorbers. A quantitative

Table 1.8.1

Coating	Reference	α_{eff} or α	ε_{eff} or ε	$\alpha_{eff}/\varepsilon_{eff}$ or α/ε
Sn doped In_2O_3	2	0.85	0.081	11
Sn doped In_2O_3	2	0.90	0.081	11
1000 Å MgF_2 180 Å TiO_2/ 180 Å Ag/ 180 Å TiO_2	2	0.54	0.017	32
Black chrome on dull nickel	63	0.923	0.085	10.9
Black chrome on bright nickel	63	0.868	0.088	9.8
Black nickel 1	63	0.877	0.066	13.3
Black nickel 2	63	0.867	0.109	8.0
Cd_2SnO_4	10	0.86	0.15	5.8

Source: Adapted from Fan and Bachner (2).

comparison between the two types of coatings can thus be made.

$$\alpha_{eff} = \frac{\left[\int_{0,25\,\mu m}^{2,5\,\mu m} T_r(\lambda)A(\lambda)d\lambda\right]}{\left[\int_{0,25\,\mu m}^{2,5\,\mu m} A(\lambda)d\lambda\right]} \tag{1.8.1}$$

$$\varepsilon_{eff} = \frac{\left[\int_{1\,\mu m}^{100\,\mu m} [1 - R(\lambda)]W_B(\lambda, T_B)d\lambda\right]}{\int_{1\,\mu m}^{100\,\mu m} W_B(\lambda, T_B)d\lambda} \tag{1.8.2}$$

where $T_r(\lambda), R(\lambda)$ = transmission and reflectivity of the heat mirror at wavelength λ, respectively, $W_B(\lambda, T_B)$ = energy distribution for a blackbody at temperature T_B, and $A(\lambda)$ = solar energy spectrum. The definition assumes that the absorber absorbs all the solar radiation transmitted by the heat mirror and also all the IR radiation reflected by the heat mirror. Table 1.8.1 compares the values of α_{eff} and ε_{eff} for some transparent conductors and α and ε for selective black absorbers.

1.9 Flat-Plate Collectors: Tabor's Concept of a Selective Surface

A flat-plate collector is the simplest of all devices used for solar energy utilization. It consists of an absorber panel which may be black or which may be a selective surface painted mildsteel, or galvanized iron (GI), or an aluminium sheet separated by a transparent cover plate. (Fig. 1.9.1). Use of

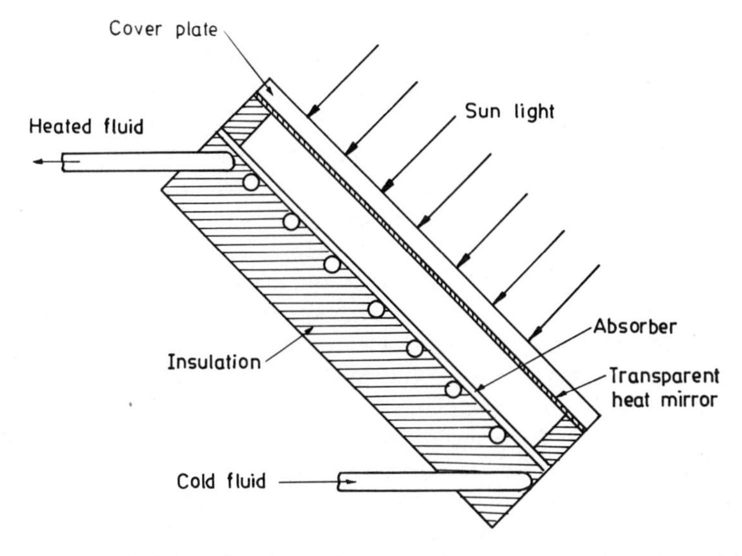

Fig. 1.9.1 Proposed design of a solar heating panel using a transparent heat mirror. Adapted from Fan and Bachner (1976) (Ref. 2).

more than one cover plates helps to reduce losses. In efficient collectors, the space between the absorber and the cover plates is evacuated. Solar radiation passes through the cover plate and is converted by the absorber into thermal energy, part of which is transferred to a heat transfer fluid. For efficient collection of solar energy the losses have to be reduced to a minimum. The losses $q_l(T)$ from a unit area of the receiver at temperature T may be written as (64):

$$q_l(T) = C(T - T_a)^{5/4} + \varepsilon_0 \sigma(T^4 - T_a{}^4) \tag{1.9.1}$$

where the first term on the right-hand side is the convection loss from a body at temperature T to the surrounding bulk air at T_a assuming the body to have an emissivity ε_0, C is a constant depending on the shape of the body and σ is the Stefan–Boltzmann constant. When the body is surrounded by an envelope of emissivity ε_1 eq. 1.9.1 becomes

$$q_l(T) = C'(T - T_1)^{5/4} + \left(\frac{1}{\varepsilon_0} + \frac{1}{\varepsilon_1} - 1\right)^{-1} \sigma(T^4 - T_1{}^4) \tag{1.9.2}$$

where T_1 is the temperature of the envelope and C' is a new constant and $\left(\frac{1}{\varepsilon_0} + \frac{1}{\varepsilon_1} - 1\right)^{-1}$ is the effective emissivity of the body. The conduction losses $q_c = b(T - T_a)$ have been assumed to be negligible. The emissivity of glass is ≈ 0.96. In eqs. 1.9.1 and 1.9.2, the radiation term is larger than the convection term for blackbodies whose temperature difference from the surroundings is small, and increases as the temperature T (where T is the temperature of the body) is raised. The losses can therefore be reduced by increasing T_1, reducing C' and reducing the effective emissivity. T_1 can be increased by using more than one cover glass, but every extra sheet causes reflection and transmission losses which reduce the value of solar energy reaching the receiver and 2–3 sheets give the best compromise. Reduction of the coefficient of convection term C' demands the evacuation of the space between the absorber and the cover plate. However, as the convection losses are much smaller compared to the radiation, the cost considerations do not demand the evacuation of the space. The possibility of reducing the effective emissivity by using the wavelength-selective absorbers and thus minimizing the radiation losses was pointed out by Tabor (64), whose contribution in this area has been very valuable and significant. Tabor centered his arguments on the basic difference which exists in the radiation from the sun and the radiation emitted by the terrestrial blackbodies. Solar radiation as received on the surface of the earth is confined to 0.3 to 2.0 μm while at 300°K the peak energy emitted by a blackbody is at 10 μm and for a body 600°K at ≈ 5 μm. In the range 300–600°K, the energy radiated in the solar region (i.e., < 2 μm) is negligible as can be seen by Planck's radiation law. The two spectra, therefore, do not overlap and this is the principle of "Wavelength discrimination" as described by Tabor (64) and Gier and Dunkle (64a). It is thus possible to reduce the effective emissivity if a surface is used in collector which has different absorption, reflection, or transmission characteristics in the solar and thermal infrared regions of the spectrum. In a reflection absorption type of filter, the selective surface should be a good absorber in the visible spectrum and a good reflector (i.e., a poor emitter) for thermal infrared. In a reflection transmission filter, the envelope will be a good reflector of thermal infrared and will transmit the solar region of the spectrum. The transparent heat mirrors come under the class of reflection–transmission type of filters while the black absorber surfaces belong to reflection–absorption type.

The evaporation of gold from a tungsten filament in an atmosphere of nitrogen with a small amount of oxygen present by Harris et al. (65) resulted in gold smoke filters which were transparent above 2 μm. Similar results on other metals were earlier reported by Pfund (66, 67). Although, the gold smoke filters gave the idea of the wavelength selectivity which can be achieved, the present work concentrates mostly on (1) light absorbing, infrared transmitting coatings produced by electrodeposition, CVD, or selective paints applied to low emittance metallic substrates, (2) conversion coatings where the metal surface is chemically converted into a compound having properties like those described under (1), (3) topological coatings, where the topology of the surface is adjusted to provide different optical properties for shortwave and longwave thermal radiation, and (4) reflection–absorption type of filters having good visible transmission and good infrared reflection.

1.10 Geometrical Spectral Selective Surfaces

Horwitz (68) introduced a new selective surface which can be fabricated by making deep holes in a metal. The holes will not be resolved by the light of larger wavelength than the hole diameter and hence will be reflected. If the hole size is properly adjusted, wavelengths shorter than 2.5 μm will be propagated down the holes as in a microwave waveguide. Fundamental waveguide theory (69) gives the onset of absorption at a wavelength λ given by $\lambda = 1.7\,d$ (or $2d$ for square guides) where d is the guide diameter. Horwitz (68) constructed a mesh type transmitting filter. Two gratings were each deposited on one side of the grooves with aluminium. The gratings were mounted face to face with the lines at right angles to each other. The result was a filter with a selective quality $\alpha/\varepsilon = 10$. Fan et al. (70) fabricated a new type of transparent heat mirror for solar energy applications by chemically etching a tin doped, In_2O_3 film to form a transparent conducting microgrid. The microgrid had openings of about 2.5 μm on a side which permitted the transmission of solar radiation. An estimate of infrared reflectivity can be obtained (4) by using a theoretical expression (71) for the reflectivity of a grid antenna made from two sets of perfectly conducting nonmagnetic parallel wires of radius r that intersect at right angles to form a square openings of side a and are joined at their intersections. If $a \gg r$, and the incident wavelength $\lambda \gg a$, the reflection coefficient is given by

$$\Omega = |R|e^{i\phi} = [1 + (\alpha k/\cos\theta)(1 - \sin^2 \tfrac{1}{2}\theta)]^{-1} \qquad (1.10.1)$$

Fig. 1.10.1 Scanning electron micrograph of microgrid fabricated from Sn doped In_2O_3 film. Adapted from Fan et al. (1976) (Ref. 70).

where $|R|$ is the reflectivity, ϕ the phase angle, $\alpha = \ln\left(\frac{a}{2\pi r}\right)$, $k = 2\pi/\lambda$, and θ is the angle of incidence from the normal. Fan et al. (70) have shown that there is a significant advantage in using transparent conductors over opaque ones. For an opaque material solar transmission, T_s is given by

$$T_s = \frac{a^2}{a^2 + \omega^2} \tag{1.10.2}$$

where ω is the linewidth. For $a = 2.5\ \mu$m, ω must be 0.2 μm in order to achieve $T_s = 0.85$. The narrow linewidth would drop the infrared reflectivity from 100 to 70%. With a transparent conductor which already has a solar transmission of 85%, the transmission can be increased to 95% by having a linewidth of 0.6 μm. Such a linewidth would decrease the infrared reflectivity of a perfect conductor by only about 1.5%. RF sputtering was used to deposit 1.35 μm thick films on Corning 7059 glass substrates from hot pressed targets of In_2O_3–15 mole %. SnO_2. The films has resistivity of 2×10^{-4} ohm-cm and had solar transmission of about 0.8. Microgrids were made by the technique of photolithography. Figure 1.10.1 gives the scanning electron micrograph of a microgrid prepared by the photolithographic technique. The optical transmission and reflectivity in the region of the solar spectrum for an unetched tin doped In_2O_3 and for a microgrid prepared from another portion of the same film are shown in the Fig. 1.10.2. The absence of the interference fringes at short wavelengths for the etched film is due to the nonuniformity of the thickness of the grid lines after etching. The performance of microgrids can be further improved by improving the conductivity of tin dopped In_2O_3 films and in this respect Cd_2SnO_4 (9, 10) coatings offer

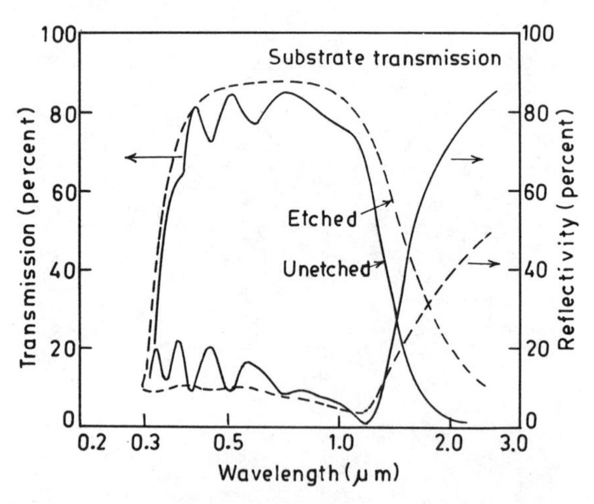

Fig. 1.10.2 Optical transmission and reflectivity of a Sn doped In_2O_3 film on glass before etching (solid lines) and after etching to form a conducting microgrid (dashed lines). Adapted from Fan et al. (1976) (Ref. 70).

promise for future applications. Infrared measurements using Gier Dunkle Reflectometer (72) were made to determine the integrated reflectivity near normal incidence in microgrids and it was found that the reflectivity is reduced by etching from 0.91 for the etched film to 0.83 for the microgrid. This decrease is much larger than the decrease calculated from eq. 1.10.1 and is probably due to the assumptions of infinite conductivity involved in deriving eq. 1.10.1.

2

Application of Selective Surfaces in Photothermal Conversion

Although the ratio α/ε is used to evaluate the selective absorbers, it has been stressed that this ratio has no significance for solar collectors. When the purpose of the absorber is to provide energy, the surface of the absorber is not under equilibrium radiation conditions, and the ratio α/ε cannot be used to evaluate the absorber. In this chapter, the efficiency of a solar collector consisting of a selective absorber and a selective transparent filter is described. The method of testing of selective surfaces under load conditions and the significant data on testing, available in the literature, is also given.

2.1 Efficiency of Transparent Filters for Solar Thermal Conversion

When an absorber is used to provide energy, e.g., in a flat-plate collector its surface will not be under equilibrium radiation conditions and α/ε is never a factor to evaluate the absorbers. Yoshida (5) derived the efficiency of a solar collector consisting of a selective absorber and a selective transparent filter. He calculated the efficiency of Drude mirror type selective transparent filters in case of a blackbody absorber and the Al_2O_3—Mo—Al_2O_3—Mo highly selective absorber. As Drude mirrors, tin doped In_2O_3 films were formed on pyrex glass plates by RF sputtering.

If the incident solar energy is Q_0, a fraction is absorbed by the absorber and the thermal radiation energy emitted by the absorber is Q_r, a fraction of which is reflected by the selective transparent filter and then reabsorbed by the absorber as Q_e (Fig. 2.1.1). Neglecting conduction and convection heat losses, the energy retained by the solar absorber (5, 6) is

$$Q = Q_a - Q_r + Q_e \tag{2.1.1}$$

and the efficiency of the collector is

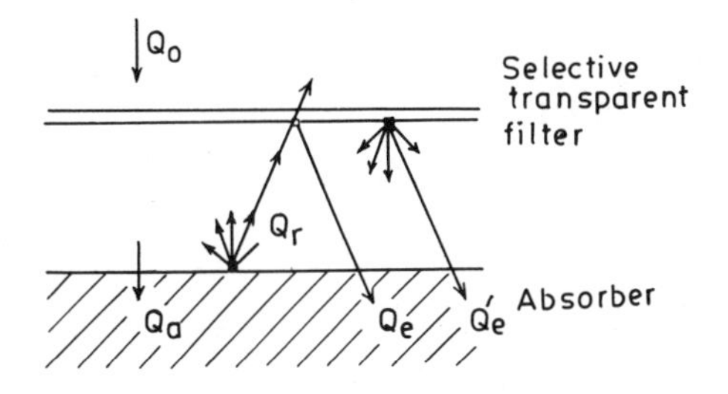

Fig. 2.1.1 Schematic diagram of a solar collector system. Adapted from Yoshida (1978) (Ref. 5).

$$\eta = \frac{Q}{Q_0} = \frac{Q_a - Q_r + Q_e}{Q_0} \tag{2.1.2}$$

The incident solar energy is given by

$$Q_0 = K \int_0^\infty A(\lambda) d\lambda \tag{2.1.3}$$

where K is the solar concentration and $A(\lambda)$ is the solar irradiation spectrum at the earth's surface. Yoshida (5) considered four types of absorbers (73, 74): (1) blackbody, (2) specular surface, (3) diffuse surfaces, and (4) general case (neither diffuse nor specular surface). The selective transparent filter has been assumed to be having specular reflection surfaces with solar irradiation normal to the surface. For a blackbody absorber we have

$$Q_a = K \int_0^\infty t'_0(\lambda) A(\lambda) d\lambda \tag{2.1.4}$$

$$Q_r = \int_0^\infty W_B(T_B, \lambda) d\lambda \tag{2.1.5}$$

$$Q_e = \int_0^\infty W_B(T_B, \lambda) \left[\frac{1}{\pi} \int_\Omega r'(\lambda;\theta,\phi) \cos\theta \, d\omega \right] d\lambda \tag{2.1.6}$$

where $t'_0(\lambda)$ and $r'(\lambda;\theta,\phi)$ are the spectral transmittance for normal incidence and the spectral reflectance for the light incident on the polar angle (θ,ϕ) of the selective transparent filter, respectively.

$r'(\lambda;\theta,\phi)$ is the average value for p and s polarized light. $W_B(T_B,\lambda)$ is the energy distribution function for blackbody radiation of the temperature T_B. The notation

$$\int_\Omega d\omega \tag{2.1.7}$$

stands for the integration over the hemispherical solid angle and

$$d\omega = \sin\theta \, d\theta \, d\phi \tag{2.1.8}$$

If the spectral reflectance is independent of the angle of incidence

$$Q_r - Q_e = \int_0^\infty W_B(T_B, \lambda)[1 - r'_0(\lambda)] d\lambda \tag{2.1.9}$$

where $r'_0(\lambda)$ is the spectral reflectance for normal incidence. Hence η can be written as follows:

$$\eta = \frac{K \int_0^\infty t'_0(\lambda) A(\lambda) d\lambda - \int_0^\infty W_B(T_B, \lambda)[1 - r'_0(\lambda)] d\lambda}{K \int_0^\infty A(\lambda) d\lambda} \tag{2.1.10}$$

This can be evaluated if the spectral reflectance and transmittance of the selective transparent filter for normal incidence are known.

If the absorber surface is not a blackbody, the multiple reflections between the selective transparent filter and the absorber (Fig. 2.1.2) should be considered and the results of such calculations are presented below (5): For absorbers having specular reflection surfaces, the total solar energy absorbed Q_a is

$$Q_a = \sum_{n=0}^{\infty} q_{an} = \int_0^\infty \frac{A(\lambda) t'_0(\lambda) \alpha_0(\lambda)}{1 - r_0(\lambda) r'_0(\lambda)} d\lambda \tag{2.1.11}$$

where $\alpha_0(\lambda)$ is the absorptance of the absorber for normal incidence, and $r_0(\lambda)$ and $r'_0(\lambda)$ are the reflectance of the absorber and the selective transparent filter, respectively, for normal incidence, q_{an} is the incident solar energy reflected n times by the absorber surface, reflected back by the filter, and then absorbed by the absorber, and is given by

$$q_{an} = K \int_0^\infty A(\lambda) t'_0(\lambda) [r_0(\lambda) r'_0(\lambda)]^n \alpha_0(\lambda) d\lambda \tag{2.1.12}$$

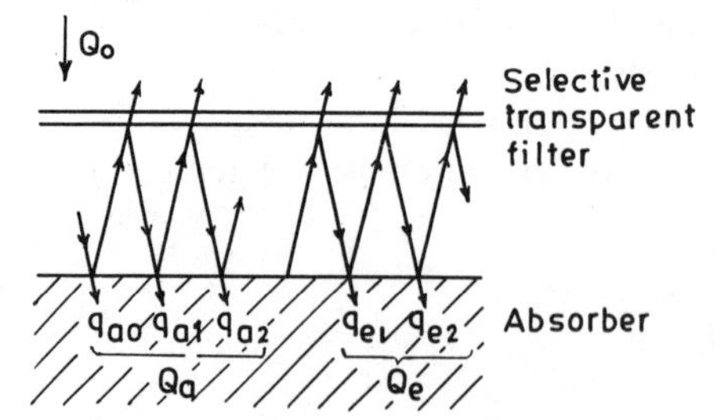

Fig. 2.1.2 Schematic diagram of the solar collector system consisting of a selective transparent filter and a non-black absorber. Adapted from Yoshida (1978) Ref. 5).

The thermal radiation energy emitted by the absorber is Yoshida (5)

$$Q_r = \int_0^\infty W_B(\lambda, T_B)\left[\frac{1}{\pi}\int_\Omega \varepsilon(\lambda;\theta,\phi)\cos\theta\, d\omega\right]d\lambda \tag{2.1.13}$$

where $\varepsilon(\lambda;\theta,\phi)$ is the spectral directional emittance of the absorber surface.

The total thermal radiation energy reabsorbed by the absorber is Q_e and is given by

$$Q_e = \sum_{n=1}^{\infty} q_{en} = \int_0^\infty W_B(\lambda, T_B)\left[\frac{1}{\pi}\int_\Omega \frac{\varepsilon(\lambda;\theta,\phi)r'(\lambda;\theta,\phi)\alpha(\lambda;\theta,\phi)}{1 - r'(\lambda;\theta,\phi)r(\lambda;\theta,\phi)}\cos\theta\, d\omega\right]d\lambda \tag{2.1.14}$$

where q_{en} is the fraction of Q_r reflected n times by the filter and then reabsorbed by the absorber, and $r'(\lambda;\theta,\phi)$ and $r(\lambda;\theta,\phi)$ are the spectral reflectance of the selective transparent filter and the absorber, respectively, for the light incident from the polar angle (θ, ϕ).

If the absorber has diffuse reflection surface, bidirectional spectral reflectivity $r(\lambda;\theta_i,\phi_i;\theta_r,\phi_r)$ is independent of the angle of reflection and is written as $r(\lambda;\theta_i,\phi_i)$, where (θ_i,ϕ_i) denotes the polar angle of incidence. Yoshida (5) has derived expressions for Q_a, Q_r and Q_e and these are given by:

$$Q_a = K\int_0^\infty A(\lambda)t_0'(\lambda)\alpha(\lambda)\left[1 + \frac{r(\lambda;0,0)R_1}{1 - R_2}\right]d\lambda \tag{2.1.15}$$

$$Q_r = \int_0^\infty W_B(\lambda, T_B)\varepsilon(\lambda)d\lambda \tag{2.1.16}$$

$$Q_e = \int_0^\infty W_B(\lambda, T_B)\varepsilon(\lambda)\alpha(\lambda)\left[\frac{R_1}{1 - R_2}\right]d\lambda \tag{2.1.17}$$

where

$$R_1 = \frac{1}{\pi}\int_\Omega r'(\lambda;\theta,\phi)\cos\theta\, d\omega \tag{2.1.18}$$

$$R_2 = \frac{1}{\pi}\int_\Omega r'(\lambda;\theta_i,\phi_i)r(\lambda;\theta_i,\phi_i)\cos\theta_i\, d\omega_i \tag{2.1.19}$$

and $\alpha(\lambda)$ and $\varepsilon(\lambda)$ are the spectral absorptance and emittance, respectively, of the absorber surface which are independent of the direction.

If the absorber surface is neither specular nor diffuse, Yoshida (5) derived expressions for Q_a, Q_r, and Q_e and they are given below:

$$Q_a = K \int_0^\infty A(\lambda) t'_0(\lambda) \left[\alpha_0(\lambda) + \sum_{n=1}^{\infty} \frac{1}{\pi} \int_\Omega r(\lambda;0,0;\theta_i,\phi_i) \right.$$
$$\left. r'(\lambda;\theta_i,\phi_i) R_{n-1}(\lambda;\theta_i,\phi_i) \cos\theta_i \, d\omega_i \right] d\lambda \qquad (2.1.20)$$

$$Q_r = \int_0^\infty W_B(\lambda, T_B) \left[\frac{1}{\pi} \int_\Omega \varepsilon(\lambda;\theta,\phi) \cos\theta \, d\omega \right] d\lambda \qquad (2.1.21)$$

$$Q_e = \int_0^\infty W_B(\lambda, T_B) \left[\sum_{n=1}^{\infty} \frac{1}{\pi} \int_\Omega \varepsilon(\lambda;\theta_i,\phi_i) r'(\lambda;\theta_i,\phi_i) \right.$$
$$\left. \times R_{n-1}(\lambda;\theta_i,\phi_i) \cos\theta_i \, d\omega_i \right] d\lambda \qquad (2.1.22)$$

where

$$R_n(\lambda;\theta_i,\phi_i) = \frac{1}{\pi} \int_\Omega r(\lambda;\theta_i,\phi_i;\theta_r,\phi_r) \times r'(\lambda;\theta_r,\phi_r)$$
$$\times R_{n-1}(\lambda;\theta_r,\phi_r) \cos\theta_r \, d\omega_r \qquad (2.1.23)$$

$$R_0(\lambda;\theta_i,\phi_i) = \alpha(\lambda;\theta_i,\phi_i) \qquad (2.1.24)$$

The above expressions for the efficiency of solar collectors in the presence of transparent coverplate and the absorber surface have been used to evaluate the efficiency of collector systems (5). Yoshida (5) used tin doped In_2O_3 (film thickness 3100 Å) on a Pyrex glass plate as a transparent filter. The optical properties of tin doped In_2O_3 were well interpreted by Drude model except below $0.4\,\mu$m, where the fundamental absorption takes place. The efficiencies were evaluated as functions of T_B and K using eq. 2.1.10. With blackbody absorber, the collector without the Drude mirror had higher efficiency than that with the Drude mirror for T_B lower than a certain temperature which depends upon K. At $T_B = 600°$ K, the collector with the Drude mirror had higher efficiency than without a Drude mirror for $K < 86$ when $E_c (= \hbar/\tau$ where τ is the relaxation time), the relaxation energy, is 0.00 eV, and for $K < 40$ where $E_c = 0.10$ eV. With highly selective absorber such as Al_2O_3—Mo—Al_2O_3—Mo (AMAM) multilayer absorber, the effective operating conditions for the use of Drude mirrors are restricted to high T_B and low K as is shown in Fig. 2.1.3. In the calculations, the CIE-recommended (75) AM1 spectrum was used by Yoshida (5).

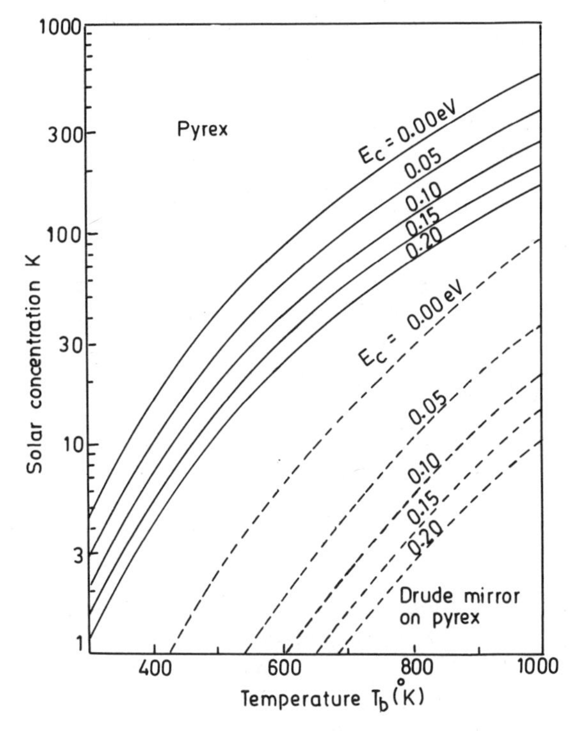

Fig. 2.1.3 Comparison of the efficiencies of the collectors consisting of the Drude mirror on a Pyrex glass plate, and that of a naked Pyrex glass plate in the case of a blackbody absorber (solid lines) and in the case of the AMAM highly selective absorber (broken lines). Adapted from Yoshida (1978) (Ref. 5).

2.2 Applications of Selective Coatings in Solar Thermoelectric Generators

Campana and Jose (76) and Sheklein (77) have described the application of solar selective surfaces in solar thermoelectric generators working without radiation concentration or with a low degree of concentration. In a thermoelectric generator the bridging plates of the hot junctions should have a high value of absorption coefficient for solar radiation and low emissivity for thermal radiations, while the emissivity of the cold junction should be close to unity with a low absorption for solar radiation. The transparent insulation of the "hot box" thermoelectric generator should form a window for solar radiation and should have good reflection characteristic for thermal infrared. Numerous works have appeared (64, 78–85) describing the material and technological aspects of selective systems. The effectiveness of selective systems is determined by comparing the thermal performance of collectors one with a selective surface and the other with a black surface. A comparative thermal analysis of the "hot box" device (Fig. 2.2.1) has been done by

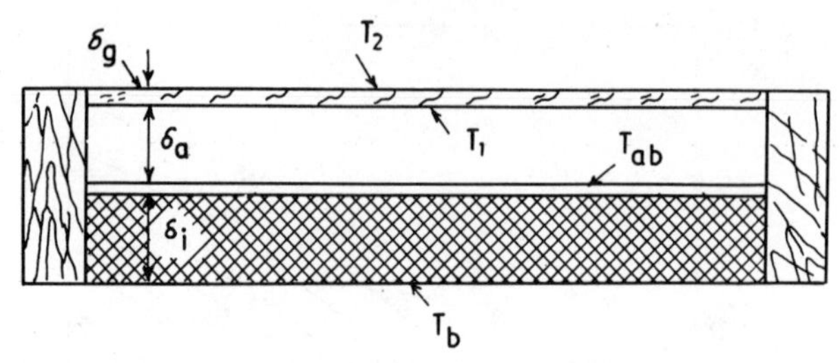

Fig. 2.2.1 Section through a solar thermoelectric generator of the "hot box" type. Adapted from Vladimirova et al. (1969) (Ref. 86).

Vladimirova et al. (86). Both selective and nonselective absorbing surfaces and transparent insulations have been considered.

Allowing for the heat losses through the bottom, the heat balance equation of a thermoelectric generator can be written as (86)

$$(T_{ab} - T_b)\frac{\lambda_i}{\delta_i} = \alpha_c(T_b - T_0) + \varepsilon_b\sigma\left[\left(\frac{T_b}{100}\right)^4 - \left(\frac{T_0}{100}\right)^4\right] \qquad (2.2.1)$$

where T_{ab} is the temperature of the absorber in °K, T_b is the temperature of the bottom of the box in °K, λ_i is the thermal conductivity of the insulation in kcal m^{-1} h^{-1} deg^{-1}, δ_i is the thickness of the insulation layer, α_c is the convective heat transfer coefficient in kcal m^{-2} h^{-1} deg^{-1}, T_0 is the temperature of the ambient air, ε_b is the emissivity of the bottom of the box, and σ is the Stefan–Boltzmann constant.

If the radiant heat losses through the bottom are not taken into consideration, we have

$$(T_{ab} - T_b)\frac{\lambda_i}{\delta_i} = \alpha_c(T_b - T_0) \qquad (2.2.2)$$

Equation 2.2.2 can be used to determine T_b at a given absorber temperature T_{ab}. In Fig. 2.2.2 the values of heat losses through the bottom (q_b) as a function of T_{ab} have been shown (86). The heat flux from the absorber to the

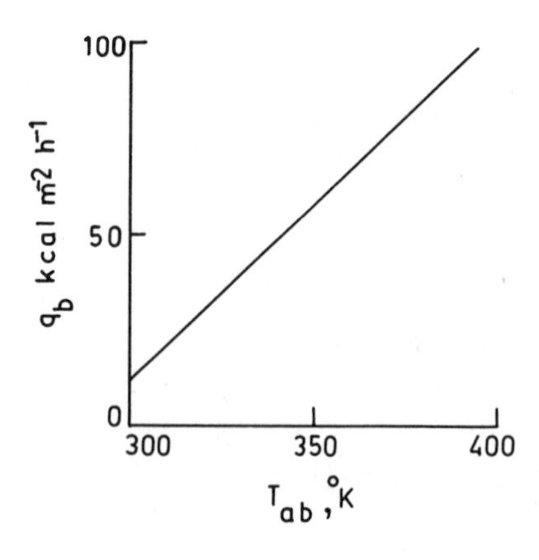

Fig. 2.2.2 Dependence of the heat losses through the bottom q_b on the absorber temperature T_{ab}. Adapted from Vladimirova et al. (1969) (Ref. 86).

inner surface of the glass q_1 is given by

$$q_1 = \frac{\lambda_{eq}}{\delta_a}(T_{ab} - T_1) + \sigma\left(\frac{1}{\varepsilon_{ab}} + \frac{1}{\varepsilon_1} - 1\right)^{-1}\left[\left(\frac{T_{ab}}{100}\right)^4 - \left(\frac{T_1}{100}\right)^4\right] \quad (2.2.3)$$

where T_1 is the temperature of the inner surface of the glass, δ_a is the air gap between the absorber and the inner surface of the glass, ε_1 is the emissivity of the inner surface of the glass, ε_{ab} is the emissivity of the absorber, and λ_{eq} is given by (87)

$$\lambda_{eq} = 0.3186(G_r P_r)^{0,25}\lambda_a \quad (2.2.4)$$

where G_r is Grashof number, P_r is Prandtl number and λ_a is the thermal conductivity of the air gap between the absorber surface and the transparent insulation glass in kcal m^{-1} h^{-1} deg^{-1}.

Grashof number, G_r is given by (88)

$$G_r = \frac{g\delta_a}{v^2}\beta(T_{ab} - T_1) \quad (2.2.5)$$

where g is the acceleration due to gravity in m/sec^2, v is the kinematic viscosity coefficient in m^2/sec. and $\beta = 1/T_{av}$, where $T_{av} = \frac{1}{2}(T_{ab} + T_1)$. After partly absorbing the solar radiation, the glass cover produces an internal source of heat (86)

$$q_{int} = EA_g \quad (2.2.6)$$

where E is the density of the total solar radiation by the glass and A_g is the coefficient of absorption of solar radiation by the glass.

The glass cover transmits

$$(T_1 - T_2)\frac{\lambda_g}{\delta_g} = q_1 T(0{\cdot}5\, q_{int}) \quad (2.2.7)$$

where λ_g is the thermal conductivity of the glass in kcal m^{-1} h^{-1} deg^{-1}, δ_g is the thickness of the glass in meters and T_2 is the temperature of the outer surface of the glass in °K.

Substituting eqs. 2.2.4, 2.2.5, 2.2.6 into equation 2.2.7 we have

$$(T_1 - T_2)\frac{\lambda_g}{\delta_g} - E\frac{A_g}{2} = \frac{0{\cdot}3186}{\delta_a^{0,25}} W(T_{ab} - T_1)^{5/4} + C_1\left[\left(\frac{T_{ab}}{100}\right)^4 - \left(\frac{T_1}{100}\right)^4\right] \quad (2.2.8)$$

where

$$W = \left(\frac{\beta P_r}{v^2}\right)^{0.25}\lambda_a = f(T_{av})$$

and

$$c_1 = \sigma\left(\frac{1}{\varepsilon_{ab}} + \frac{1}{\varepsilon_1} - 1\right)^{-1}$$

The dependence $W = f(T_{av})$ is shown in Fig. 2.2.3. Heat absorbed in the glass

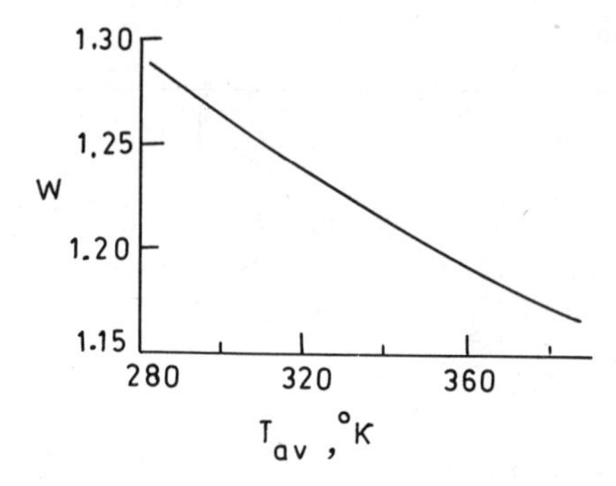

Fig. 2.2.3 Dependence of the function W on T_{av}. Adapted from Vladimirova et al. (1969) (Ref. 86).

cover is

$$\alpha_c(T_2 - T_0) + \varepsilon_2 \sigma\left[\left(\frac{T_2}{100}\right)^4 - \left(\frac{T_0}{100}\right)^4\right] = q_1 + q_{\text{int}} = (T_1 - T_2)\frac{\lambda_g}{\delta_g} + \frac{EA_g}{2} \tag{2.2.9}$$

where ε_2 is the emissivity of the outer surface of the glass. Equation 2.2.9 was used to determine T_1 (86). Putting $\lambda_g/\delta_g = a$ and $\varepsilon_2 \sigma = c_2$ we have

$$T_1 = T_2\left(1 + \frac{\alpha_c}{a}\right) + \frac{c_2}{a}\left(\frac{T_2}{100}\right)^4 - \left[\frac{\alpha_c}{a} T_0 + \frac{c_2}{a}\left(\frac{T_0}{100}\right)^4\right] - \frac{EA_g}{2a} \tag{2.2.10}$$

Writing

$$1 + \frac{\alpha_c}{a} = m, \quad \frac{c_2}{a} = n, \frac{\alpha_c}{a} T_0 + \frac{c_2}{a}\left(\frac{T_0}{100}\right)^4 = p, \frac{EA_g}{2a} = r$$

Eq. (2.2.9) becomes

$$T_1 = m\,T_2 + n\left(\frac{T_2}{100}\right)^4 - p - r \tag{2.2.11}$$

Table 2.2.1 Results of Calculations of T_1 and T_2 (in °K)

T_2	$T_1 - T_2$ For A_g equal to			T_2	$T_1 - T_2$ For A_g equal to		
	0.14	0.05	0.06		0.14	0.05	0.06
290	– 0.026	0.077	0.053	315	1.702	1.805	1.781
295	0.311	0.414	0.390	330	2.792	2.894	2.871
300	0.653	0.756	0.732	350	4.313	4.415	4.392
305	0.998	1.101	1.077	380	6.014	6.116	6.093

Source: Adapted from Vladimirova et al. (86).

Equation 2.2.8 can be represented in the form

$$(T_1 - T_2)a = \frac{0{\cdot}3186}{\delta_a^{0.25}} W(T_{ab} - T_1)^{5/4} - c_1\left[\left(\frac{T_{ab}}{100}\right)^4 - \left(\frac{T_1}{100}\right)^4\right] = \tfrac{1}{2}EA_g \qquad (2.2.12)$$

The results of calculations of T_1 and T_2 (°K) are given in Table 2.2.1 (adopted from Ref. 86).

Knowing the values of T_1 and T_2, one can find the heat losses using eq. 2.2.9. The total heat lost

$$q_{\text{lost}} = q_g + q_b \qquad (2.2.13)$$

where q_g is the heat lost through the glass cover and q_b is the heat lost through the bottom of the box.

$$q_{\text{lost}} = \frac{\lambda_g}{\delta_g}(T_1 - T_2) + \frac{EA_g}{2} + q_b \qquad (2.2.14)$$

2.3 Efficiency of Thermoelectric Generators

The transmission of glass is given by

$$E(1 - R_g - A_g) = ED_g \qquad (2.3.1)$$

where E is the density of solar radiation incident on the glass in kcal m^{-2} h^{-1}, R_g is the coefficient of reflection of solar radiation by the glass, A_g is the coefficient of absorption of solar radiation by the absorber, and D_g is the coefficient of transmission of solar radiation by the glass.

The absorber absorbs q_{ab}:

$$q_{ab} = ED_g A_{ab} \qquad (2.3.2)$$

The useful energy is q_u:

$$q_u = q_{ab} - q_{\text{lost}} = ED_g A_{ab} - q_{\text{lost}}$$

Hence the efficiency η is

$$\eta = \frac{q_u}{E} = D_g A_{ab} - \frac{q_{\text{lost}}}{E} \qquad (2.3.3)$$

Results of thermal calculations by Vladimirova et al. (86) are shown in Table 2.3.1 and Fig. 2.3.1.

In the calculations, the absorption coefficient for ordinary window glass was taken to be $A_g = 0.05$ and for selective glass, $A_g = 0.06$ and 0.14, respectively. For a sheet iron absorber, $\varepsilon_{ab} = 0.937$ and for selective absorber $\varepsilon_{ab} = 0.2$ and 0.3, respectively. For ordinary glass $R_g = 0.08$ and for selective glass $R_g = 0.12$. The other values used in the calculations were as follows: $T_0 = 288$°K; $\delta_a = 0.04$ m, $\delta_g = 0.0021$ m; $\delta_i = 0.04$ m, $\lambda_g = 0.64$ kcal

Table 2.3.1

Serial No.	T_{ab}	T_2	T_1	Glass	A_g	ε_{ab}	q_b	q_u	η
1	303	294.1	294.31	Selective	0.14	0.300	14	141	0.608
2	303	292.8	293.05	Ordinary	0.05	0.937	14	108	0.690
3	303	291.6	291.75	Selective	0.06	0.200	14	81	0.695
4	315	293.4	293.65	Selective	0.06	0.200	24	134	0.620
5	315	295.8	296.27	Ordinary	0.05	0.937	24	184	0.584
6	315	295.8	296.17	Selective	0.14	0.300	24	184	0.548
7	340	297.4	297.96	Selective	0.06	0.200	49	244	0.460
8	340	303.2	303.79	Ordinary	0.05	0.937	49	369	0.320
9	340	300.2	301.88	Selective	0.14	0.300	49	304	0.375
10	390	311.3	312.75	Selective	0.14	0.300	96	591	− 0.035
11	390	306.6	307.79	Selective	0.06	0.200	96	491	0.110
12	390	321.8	324.04	Ordinary	0.05	0.937	96	816	− 0.320

Source: Adapted from Vladimirova et al. (86).

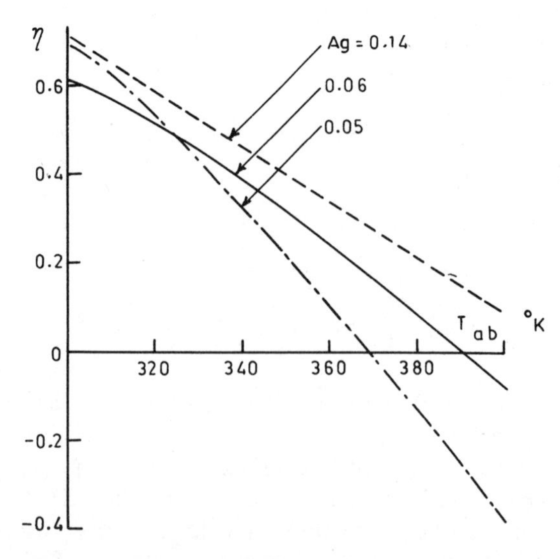

Fig. 2.3.1 Dependence of the efficiency, η, of the generator on the absorber temperature T_{ab} for various values of the absorption coefficients of glass, A_g. Adapted from Vladimirova et al. (1969) (Ref. 86).

$m^{-1}\ h^{-1}\ deg^{-1}$; $\lambda_i = 0.04$ kcal $m^{-1}\ h^{-1}\ deg^{-1}$; $E = 700$ kcal $m^{-2}\ h^{-1}$; $\sigma = 4.88$ kcal $m^{-2}\ h^{-1}\ deg^{-1}$; $\alpha_c = 16$ kcal $m^{-2}\ h^{-1}\ deg^{-1}$; $\varepsilon_1 = \varepsilon_2 = 0.937$; $A_{ab} = 0.92$.

It was concluded that the selective transparent insulation increased the efficiency and that this insulation is particularly useful at working temperatures above about 330°K. At low working temperatures the selective insulation does not contribute to efficiency as the total loss reduction due to the use of such insulation is less than the energy absorbed in the insulation.

2.4 Tabor's Calculations

Tabor (64) has given the results of theoretical calculations for flat-plate and cylindrical selective and nonselective receivers without and with partial vacuum. The flat-plate nonselective receiver using one cover glass and operating an ideal Carnot engine was found to have an optimum theoretical overall efficiency of 6.2% at a working temperature of about 66°C under peak sunshine conditions of Jerusalem. With two cover glasses, the theoretical optimum efficiency for peak sunshine is 7.9% at 86°C or about 5.2% at 68°C for the yearly average. For a cylindrical receiver about 15 cm in diameter placed at the focus of a parabolic mirror 300 cm wide (concentrating power $P = 300/15\pi = 6.37$), the theoretical optimum efficiency under peak sunshine conditions is 14.4% at 150°C or 13.6% at 143°C for the optimum for the whole year, assuming a heliostatic mounting. The extent to which the efficiencies improve in selective receivers without and with partial vacuum are given in the Tables 2.4.1 and 2.4.2 for flat-plate and cylindrical receivers

Table 2.4.1 Flat-Plate Receiver, One Cover Glass

	Nonselective Receiver		Selective Receiver		
	Without Vacuum	Partial Vacuum	Without Vacuum	Partial Vacuum	Partial Vacuum
Emissivity	0.95	0.95	0.1	0.1	0.05
Optimum operating temperature for solar input of 1.05 kw/m^2 (°C)	66	76	112	177	238
Theoretical efficiency with Carnot engine (%)	6.2	7.2	11.5	18.6	21.8
Optimum temperature for whole year (°C)	55	63	91	138	173
Theoretical efficiency (%)	4.2	5.0	8.0	13.9	16.9

Source: Adapted from Tabor (64).

Table 2.4.2 Cylindrical Receiver, Concentrating Power 6.37

	Nonselective Receiver Emissivity = 0.95	Selective Receiver Emissivity = 0.1
Optimum operating temperature for solar input of 0.95 KW/m^2 (°C)	150	271
Theoretical efficiency of Carnot cycle (%)	14.4	22.8
Optimum temperature for whole year (°C)	143	254
Theoretical efficiency (%)	13.7	21.8

Source: Adapted from Tabor (64).

respectively. The selective receivers are found to give much more satisfactory performance.

2.5 The Heat Mirrors as an Alternative to the Selective Absorber

Winegarner (8) presented the results of theoretical calculations and experimental data on the efficiencies of two cover flat-plate collectors in three modes of operation namely (1) with two cover glass plates and the flat black absorber, (2) with two cover glass plates and the selective absorber, and (3) with a heat mirror on the inside cover and flat black absorber surface. Figure 2.5.1 is a schematic diagram showing the energy and temperature distributions within various two cover glass flat-plate collectors. Collector A has a black absorber surface ($\alpha = 0.92$, $\varepsilon = 0.92$). Collector B has a selective

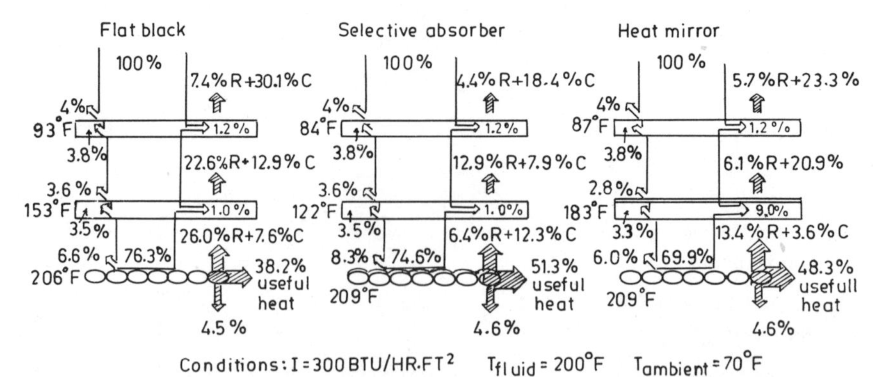

Fig. 2.5.1 Comparison of flat-plate collector designs. Adapted from Winegarner (1976) (Ref. 8).

absorber surface with $\alpha = 0.9$, $\varepsilon = 0.10$. Collector C has a black absorber surface as in A, but the outer side of the inner glass cover is a coated with a heat mirror. Absorption and reflection losses at the cover glasses and absorber surface reduce the amount of solar energy absorbed by the plate to 76.3%, approximately half of which is absorbed by the working fluid after conduction, convection, and radiation losses. When the black absorber is replaced by a selective absorber, its low infrared emittance reduces the radiation losses from 26.4 to 6.4% which goes to the working fluid increasing the useful heat collected from 38.2 to 51.3% as shown in collector B. When a heat mirror is used on the outer side of the inner glass cover with a flat black absorber surface, the radiation losses are reduced due to the reflection of the thermal infrared emitted by the absorber. However, the heat mirror absorbs about 9% of the incident energy, thus increasing its temperature. The temperature of the inner pane is 183°F as compared to 153°F for collector A and 122°F for collector B. Thus the temperature difference between the absorber plate and the inner pane is reduced, thereby reducing the temperature and radiation losses. The losses for A are 26.0 R + 7.6 C = 33.6% and are reduced in collector C to 13.4 R + 3.6 C = 17.0%. With the reduction in losses from the absorber plate, the useful heat collected is raised from 38.2 to 48.3%. The improvement in heat collected by using a heat mirror is thus comparable to that achieved by using a selective absorber.

Energy and temperature distributions in a flat-plate collector involves the solutions of nonlinear integral–differential equations which are difficult to solve. Simplified solutions can however be obtained by replacing the real continuous system by a series of discrete isothermal elements, which can be coupled through thermal resistances from heat transfer laws. As the heat and direct current flows are analogous, direct current theory can be applied to heat transfer in a collector. An isothermal mass is considered a node and assuming a discontinuous temperature variation from node to node the heat balance equation for node i is written as

$$(MC_p)_i \frac{dT}{d\theta} = \sum K_{iJ}(T_J{}^4 - T_i{}^4) + \sum \frac{T_J - T_i}{R_{iJ}} + A_i I(\theta)\alpha \qquad (2.5.1)$$

where M is the thermal mass, R is the thermal resistance, C_p is the heat capacity, I is the solar flux, A is the area, θ is the time, K is the radiation coefficient, α is the solar absorptance, and T is the absolute temperature.

The temperature of each node is dependent on all the other nodes in the system. The temperature of a particular node can be obtained only from the solution of N simultaneous equations. Computer solutions can be obtained by finite difference methods (8). Figure 2.5.2 shows the plots of the instantaneous collector efficiency for three collectors as a function of excess temperature (plate temperature–air temperature) for three insolution levels: 300, 200 and 100 BTU/hr ft^2 taken from Winegarner's data (8).

In the temperature range 40°F to 100°F, heat mirror and selective absorber designs are equivalent and are better than flat black absorber design. Wine-

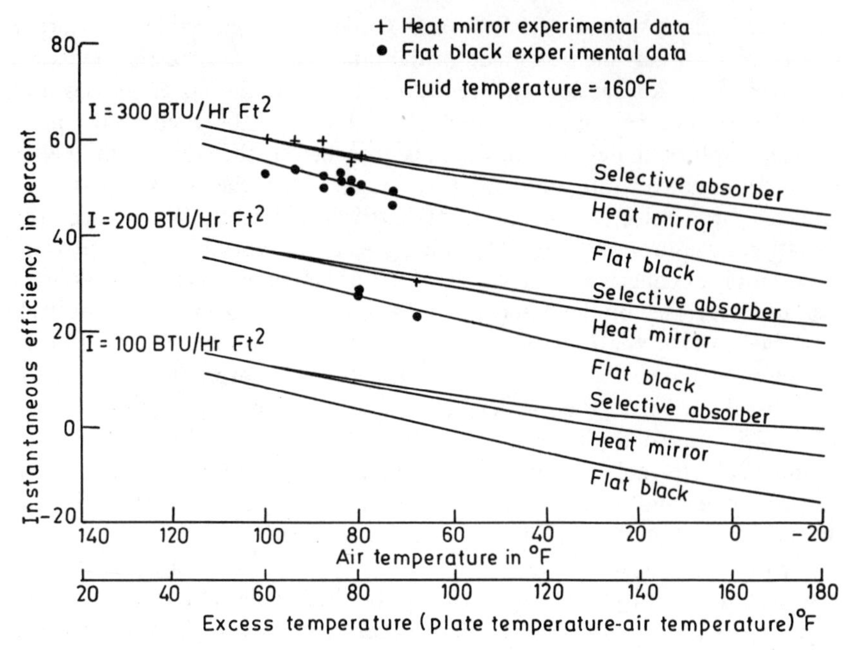

Fig. 2.5.2 Instantaneous efficiency in percent as a function of excess temperature (plate temperature – air temperature) (°F). Adapted from Winegarner (1976) (Ref. 8).

garner (8) calculated daily efficiency differences by plotting heat absorbed with time by three collectors. Comparison of the areas under these curves with the area under the solar insolation curve gave the daily efficiency of each collector. The results are shown in Fig. 2.5.3. These experiments confirmed that a heat mirror collector is nearly as good as the selective absorber collector and both are of course superior to standard flat black collector.

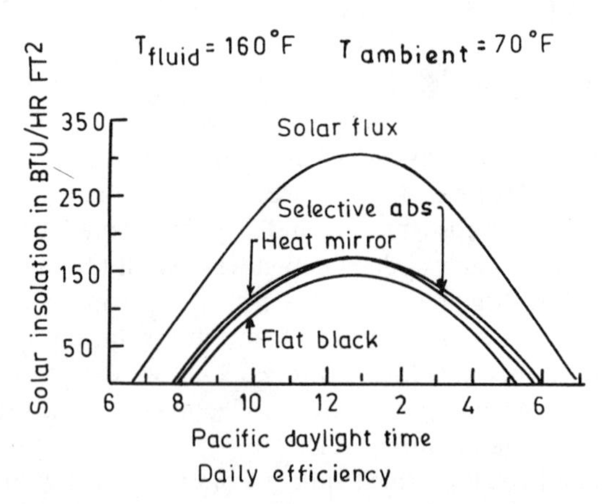

Fig. 2.5.3 Heat absorbed versus time by each of the three collector designs during a July day in Santa Rosa, Ca. Adapted from Winegarner (1976) (Ref. 8).

Experimental data on heat mirror and black absorber designs were obtained (8) in a collector test facility which was a comparative device. Two collectors were connected to heat storage system with water as a heat transfer fluid with common fluid input and separate output lines. Comparative data was obtained by measuring the flow rate through each collector and the temperature differences in the input and output fluid. In order to measure temperature differences, the water from a common input and two output lines was periodically sampled by valves over to a thermocouple, whose signal was amplified and recorded. This technique not only gives the temperature difference but also temperature drift in the system. Figure 2.5.2 gives the experimental data points superimposed on to theoretical curves. The heat mirror data points are indicated by plus signs and the black absorber data points by circles. Winegarner's experimental results were in agreement with theoretical calculations.

Winegarner (8a) did model calculations of the effect of using a selective absorber compared to a nonselective one for a concentrating parabolic collector at 315°C (600°F) with ambient temperature of 21°C (70°F). The

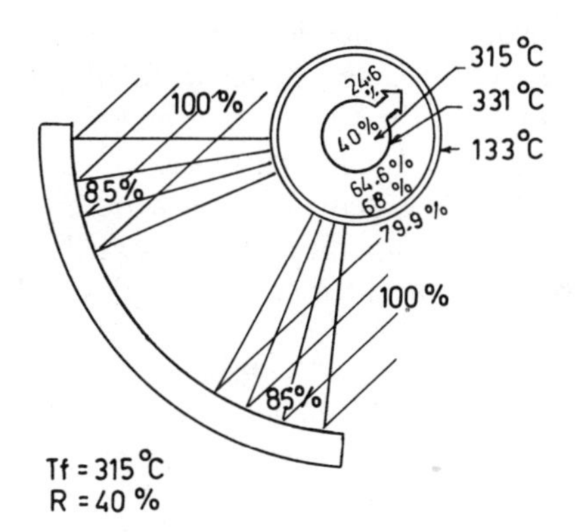

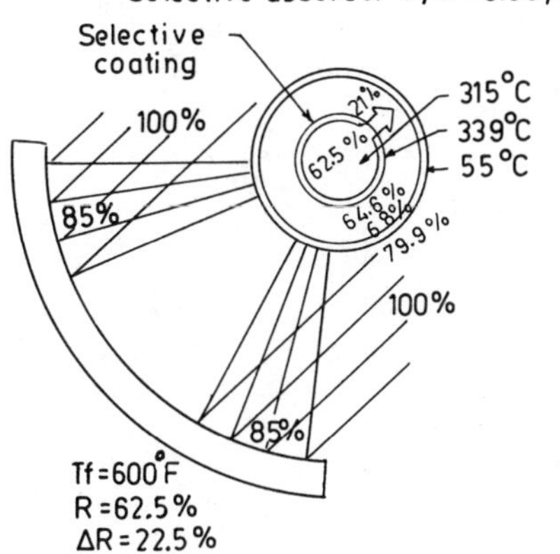

Fig. 2.5.4 The effect of using a selective absorber compared to a nonselective for a concentrating parabolic collector operating at 315°C (600°F) with ambient temperature of 21°C (70°F). Notation (R) Net collection efficiency; (ΔR) change in collection efficiency over that of a standard black absorber. Adapted from Winegarner (1975) (Ref. 8*a*).

results are shown in Fig. 2.5.4. At higher operating temperatures a selective surface becomes a very important consideration. The net efficiency can increase from 12–22% for a parabolic collector and collection efficiencies as high as 45.6 and 62.5% can be achieved by the use of selective absorbers.

Goodman and Menke (89) studied the effect of the change in the cover plate on solar collector efficiency. Their work covered the following areas: (1) change in the efficiency of a flat-plate collector with a selective or nonselective coating by coating the cover plate with a metal oxide, (2) the effect on collector efficiency using an evacuated cover plate that has spacers between the plates, and (3) the effect on the efficiency of a flat-plate collector using low-iron and iron dereflecting glass. He found that the heat mirror coating offers some improvement in a flat-black two-cover plate collector but no improvement in a selective black one-cover plate collector. Low-iron and dereflecting soda lime glass gives considerable improvement over regular soda lime glass. Evacuated single cover selective black gives 50% efficiency at 430°F. Support studs of 5% cross-sectional area of the collector appreciably lower the efficiency of the unit. The collector plate with heat mirror and low-iron glass attains higher equilibrium temperature under no load conditions.

Goodman and Menke (89) obtained experimental data on no-load equilibrium temperatures for nonselective collector plates with different combination of cover plates. Table 2.5.1 gives the various combinations by cover and collector plates, and the experimental temperatures are recorded next to predicted temperatures in parentheses. Calculated values are higher than experimental values. Higher no-load temperatures for coated and low iron glass are obtained as compared to uncoated soda lime glass. The

Table 2.5.1 Black with $\varepsilon = 0.96$, Covered with Two 0.1 in. Glass Plates (3rd or 4th Surface Refers to Fig. 1, Ref. 89; Numbers in Parentheses are Predicted Values in °F)

Resistance of SnO_2 Coatings (ohms/Sq.)	SnO_2 on 3rd Surface	SnO_2 on 4th Surface	Low Iron Glass SnO_2 on 3rd Surface	Low Iron Glass SnO_2 on 4th Surface
No coating	238(268)		257(283)	
31	244(289)	234(292)	—	—
40	256(287)	254(291)	270(308)	—
56	—	—	265(303)	—
62	254(281)	258(286)	262(301)	256(306)
78	253(278)	252(282)	265	—
94	250(274)	245(277)	—	—

Source: Adapted from Goodman and Menke (89).

31 ohm/sq. coatings applied to No. 4 surface of the glass gives an equilibrium plate temperature of 234°F, which is less than with an uncoated glass. This result is rather unexpected. No data on efficiencies under load conditions and with selective and nonselective coatings with coated and uncoated cover plates have been reported by the authors (89).

2.6 Testing of Selective Surfaces in a Flat-Plate Collector under Load Conditions

The useful energy available per unit time per unit area, $\dot{Q}''_u$ in a flat-plate collector is given by (73):

$$\dot{Q}''_u = H_T\left[\eta_0 - U_L\frac{(\bar{T}_\omega - T_a)}{H_T}\right]F_p \tag{2.6.1}$$

where H_T is the solar insolation on a tilted surface, η_0 is the optical efficiency ($=\alpha T$, where α is the absorption coefficient and T is the transmission), U_L is the heat loss coefficient, $\bar{T}_p$ is the average plate temperature, T_a is the ambient temperature, $\bar{T}_\omega$ is the average water temperature [$=\frac{1}{2}(t_0 + t_i)$ where t_i is the inlet and t_0 outlet temperature in a flat-plate collector],

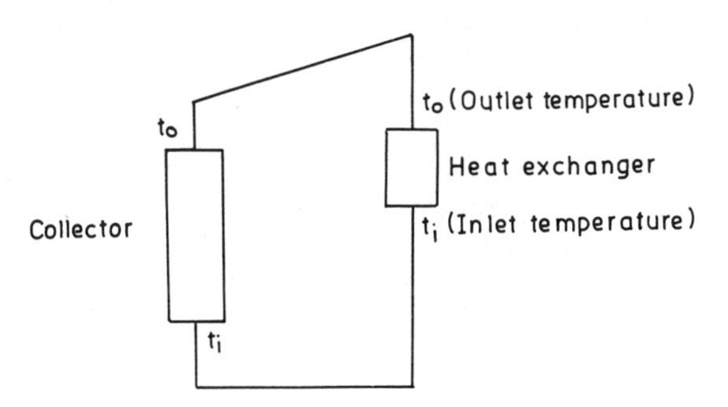

Fig. 2.6.1 Block diagram of a collector with heat exchanger.

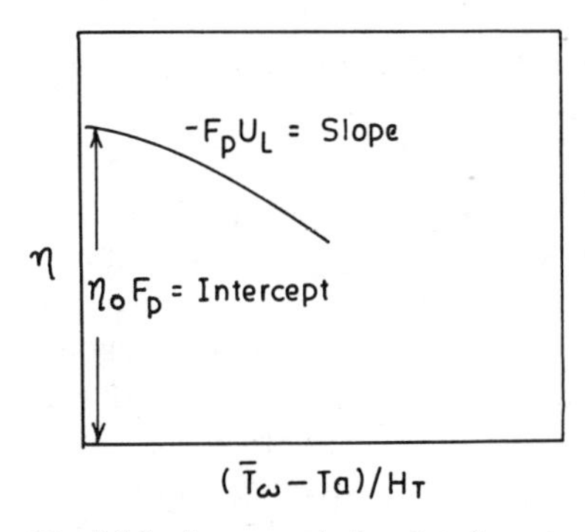

Fig. 2.6.2 A representative plot of η as a function of $(\bar{T}_\omega - T_a)/H_T$.

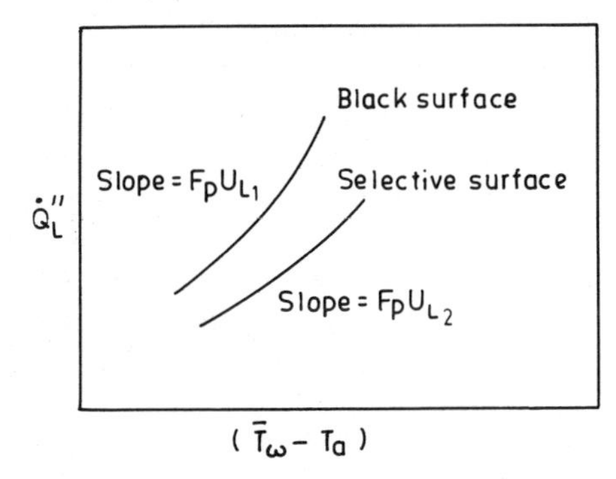

Fig. 2.6.3 A representative plot of $\dot{Q}''_l$ as a function of $(\bar{T}_\omega - \bar{T}_a)$.

F_p is the plate efficiency factor, $\eta(=\dot{Q}_u''/H_T)$ is the collector efficiency, and $H_T\eta_0$ is the energy absorbed.

If $\dot{m}$ is the mass of water flowing per unit time and C_p is the specific heat, we have

$$\dot{Q}_u'' = \dot{m}\,C_p(t_0 - t_i) \tag{2.6.2}$$

In terms of efficiency, η, eq. 2.6.1 can be written as

$$\eta = \left[\eta_0 - U_L \frac{(\bar{T}_\omega - T_a)}{H_T}\right] F_p \tag{2.6.3}$$

$$\eta = \eta_0 F_p - \frac{U_L F_p(\bar{T}_\omega - T_a)}{H_T} \tag{2.6.4}$$

If one therefore plots a graph of between η versus $(\bar{T}_\omega - T_a)/H_T$, the slope will give $-F_p U_L$ and the intercept with η axis will give $\eta_0 F_p$. If η_0 is known, F_p can be determined.

If $H_T = 0$, the energy will be lost. Writing $\dot{Q}_L''$ as the energy lost per unit time per unit area, we have

$$\dot{Q}_L'' = -U_L(T_\omega - T_a)F_p \tag{2.6.5}$$

The energy lost per unit area per unit time can be calculated by knowing the mass of water flowing through the collector per unit time and the specific heat and inlet and outlet temperatures:

$$\dot{Q}_L'' = \dot{m}\,C_p(t_i - t_0) \tag{2.6.6}$$

In the case when the solar insolation is zero, the inlet temperature will be more than the outlet temperature. The experiment is performed in the dark and a heater is used to change $(\bar{T}_\omega - T_a)$. A plot of $\dot{Q}_L''$ versus $(\bar{T}_\omega - T_a)$ gives nearly a straight line, the slope of which gives $F_p U_L$. Since F_p has already been determined in the first part of the experiment, the heat loss coefficient can be determined. If the experiment is therefore performed first with a black surface having a heat loss coefficient U_{L_1}, and then with a selective surface having a heat loss coefficient U_{L_2}, the heat losses in the two cases can be compared.

In Fig. 2.6.1, a block diagram of the collector with a heat exchanger with t_i the inlet temperature and t_0 the outlet temperature is shown. In Fig. 2.6.2 the plot of η as a function of $(\bar{T}_\omega - T_a)/H_T$ is shown. The slope gives $-F_p U_L$ and the intercept gives $\eta_0 F_p$. Knowing η_0 from the basic properties of the surface, F_p is evaluated. Figure 2.6.3 gives a plot of $\dot{Q}_L''$ as a function of $(\bar{T}_\omega - T_a)$. The slope gives $F_p U_L$. Knowing the value of F_p, the heat loss coefficients U_{L_1} and U_{L_2} for a black and selective surfaces respectively are compared. Nandwani et al. (90) have described the comparative testing of selective black nickel coated solar collector against one with black board paint. The nickel black selective coating was found to produce substantial improvement in performance of a solar collector with single glass cover, specially at higher temperatures and/or at lower levels of insolation.

3

Transparent Conductors in Photovoltaic Energy Conversion

There has been a considerable interest during the last few years in the use of transparent and conductive layers as a constitutive element of a photovoltaic structure (92–114). Such devices are of current interest because they represent a potentially low-cost method for fabricating large-scale solar energy conversion arrays. Although pure single crystal indium and tin oxides are insulators, polycrystalline samples exhibit high *n*-type conductivity, believed to be caused by oxygen vacancies and substitutional impurities. When deposited upon silicon, indium phosphide, and gallium arsenide, indium and tin oxides form good photovoltaic cells. The cost reduction of such cells is in the junction formation step and also in eliminating the antireflection layer. Basically In_2O_3, SnO_2, or tin doped In_2O_3 (ITO) are deposited onto single crystal silicon substrates by a technique such as electron beam or spray. These processes are a low-temperature step and could reduce degradation of carrier lifetimes, a problem observed in some high-temperature diffusion processes. Compared to conventional silicon cells, the oxide semiconductor on silicon exhibits the following advantages: (1) the oxides exhibit indices of refraction in the right range to provide an inherent antireflection coating on silicon (2) the carrier generation occurs directly in the depletion region, increasing the quantum yield towards the short wavelength region, (3) the high optical transparency and low electrical resistivity greatly simplify making front contact, (4) the effective zero depth junction eliminates any surface dead layer and improves the response at short wavelengths, and (5) the oxides absorb in the ultraviolet and therefore act as a window for sunlight. Moreover, in such cells as the base doping increases, the dark current decreases, allowing the lifetime, collection depth, radiation resistance, and efficiency to increase. The ion beam sputtering and spray techniques used to make diodes are low-cost processes.

3.1 Interfacial Layer Heterojunctions

In the Schottky barrier devices, the presence of an oxide layer between the metal and semiconductor increases considerably the conversion efficiency. The interfacial SiO_2 layer is present because of the following two reasons: (1) silicon is exposed to the air before the deposition of ITO, SnO_2, or In_2O_3, and (2) the reactions of both In_2O_3 and SnO_2 with silicon yield SiO_2 exothermically. Stirn and Yeh (96) have reported that the addition of an oxide layer about 30–50 Å on GaAs can yield devices with about 15% efficiency. In evaluating the performance of ITO or In_2O_3 or SnO_2/Si cells, the theory of metal–insulator–semiconductor (MIS) devices is useful (97). MIS structure offers a means of overcoming the principal deficiency of Schottky barrier solar cells, namely low open circuit voltage, while maintaining the attractive features that have led the metal semiconductor junction to be considered as a possible alternative to the *p–n* junction for large area, terrestrial solar applications (Pulfrey, Ref. 99a). In such structures the thermionic emission dark current can be reduced by either increasing the effective metal–semiconductor barrier height, decreasing the probability of majority carrier tunneling, encouraging interface states with a large capture cross section for majority carriers, or reducing the number of majority carriers at the semiconductor surface.

Although the role of interfacial layer is not clearly understood, several attempts have been made to explain this. Fonash (98) has proposed that an interfacial insulating or semiconducting layer can be used to improve the open circuit voltage and also the light-generated carrier response by field shaping. For a given light-generated current, one would like to obtain the smallest value possible at a given total voltage for the two bucking terms—the thermionic emission term and the diffusion term. According to Shewchun et al. (99), significant tunnel currents can flow between the metal and the semiconductor when the insulator is sufficiently thin. The operation of the cell is based on the concept that they are minority carrier nonequilibrium MIS tunnel diodes, according to these authors. They have presented efficiency as a function of insulator thickness, substrate carrier concentration, surface states and oxide charge. For an insulator thickness of about 10 Å, the cell exhibits Schottky diode behavior, and a maximum theoretical efficiency of 21% is possible under AM2 illumination for high substrate doping and low interface defect density.

ITO or antimony doped SnO_2 are degenerate semiconductors and behave essentially as metallic electrodes where the contact with the base semiconductor is through the conduction band. Various combinations of oxide semiconductors and base materials have been reported in the literature. While silicon has been investigated in most detail (91, 103, 104, 113, 115–119), Ge (105, 113), CdTe (120), InP (113, 120, 121), GaAs (105, 113), and $CuInSe_2$ (122) have received some attention. The crystal structures, lattice parameters, and thermal expansion coefficients of the device components do not match (101) and yet it is very intriguing that some of the photovoltaic structures have high efficiencies of more than 10% (91, 97, 119). Silicon has a diamond structure with a lattice parameter $a = 0.5431$ nm, SnO_2 is tetragonal with lattice parameters $a = 0.4737$nm and $c = 0.3185$ nm, and In_2O_3 is cubic with $a = 1.0118$ nm (123). The compounds SnO_2, In_2O_3, or ITO are therefore completely uncompatible with silicon and the devices SnO_2, In_2O_3, ITO/Si are expected to have high degree of disorder at the interface with poor conversion efficiency. The performance and high efficiency of such devices has been explained on the basis of a semiconductor–insulator–semiconductor (SIS) model (101). The solar cell has been termed as "nonequilibrium minority carrier tunnel current transport" device. The wide band gap semiconductor blocks band-to-band majority carrier currents and thus an SIS cell gives better performance than MIS cell. The effects of interfacial layer thickness, substrate doping level, surface states, and interface charge and temperature on the performance of SIS devices has been predicted (101) and an efficiency of 20% for ITO/Si has been contemplated.

3.2 SIS Model

A satisfactory theory to explain the ITO/Si type of system has been given by Shewchun et al. (101) and is based upon the behavior of a thin insulating layer between ITO and Si. This theory could satisfactorily explain (124) high efficiencies obtained in ITO–InP solar cells (121). The MIS diode has been shown to be a limiting case of SIS theory. Since the nature of the interface layer is not clearly understood, the theory assumes a "sharp interface layer" model. The interface layer has been assumed to be sufficiently thin that the current transport is by tunneling. The energy band diagram of ITO–SiO_2–p-Si is shown in Fig. 3.2.1. The interfacial layer (10–20 Å) is that of a wide gap semiconductor or insulator. The layer is formed during the device fabrication and its nature may be more complicated than assumed, due to the entrance of components from ITO (101). A consideration of the ITO and Si work functions has shown that the surface of the silicon will be inverted. The large band gap of ITO prevents the holes from p-silicon from band to band tunneling. The inverted p-silicon surface provides minority carriers which tunnel into ITO (J_{CT}). J_{ST} is the tunnel current via defects in ITO–Si, J_{C1} and J_{V1} are the effective coupling current flows due to interchange of charge between the conduction and the valence bands of the silicon by recombination generation. With decreasing thickness of the insulator, the tunnel currents increase and below a critical thickness the diodes operate as "nonequilibrium" tunnel SIS diodes. In a certain bias range, the diodes operate in the nonequilibrium mode where the current is semiconductor limited. The ITO–Si cell operates as a minority carrier nonequilibrium SIS tunnel diode where the minority carrier quasi-Fermi levels in the

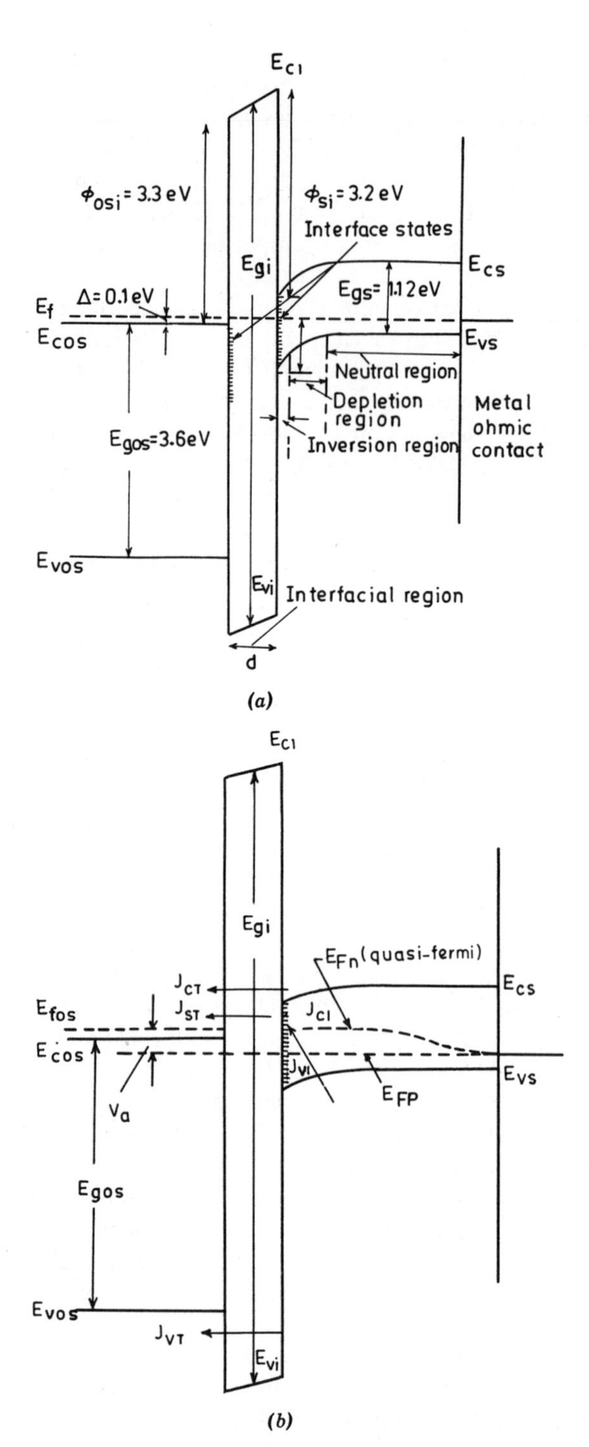

E_cI
φ_osi = 3.3 eV
φ_si = 3.2 eV
Interface states
E_gi
E_cs
Δ = 0.1 eV
E_gs = 1.12 eV
E_f
E_cos
E_vs
Neutral region
Depletion region
Metal ohmic contact
E_gos = 3.6 eV
Inversion region
E_vos
E_vi
Interfacial region
d
(a)
E_cI
E_gi
E_Fn (quasi-fermi)
J_CT
E_cs
J_ST
J_cI
E_fos
E_cos
J_vI
E_vs
E_FP
V_a
E_gos
E_vos
J_VT
E_vi
(b)

Fig. 3.2.1

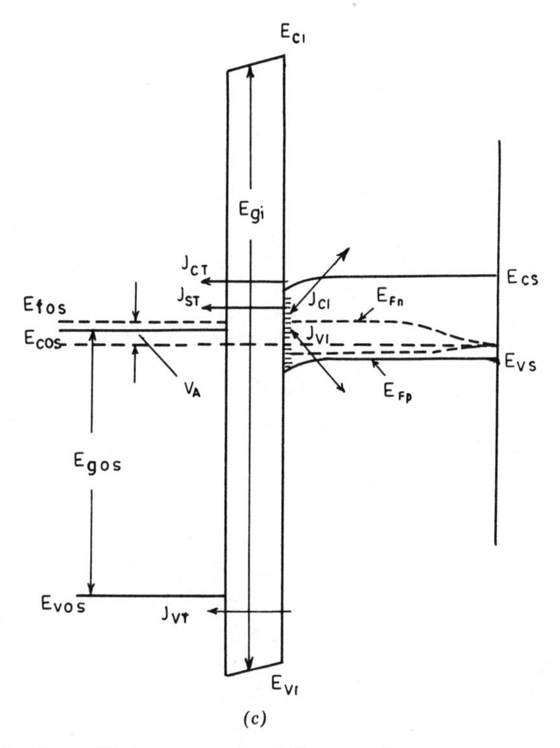

(c)

Fig. 3.2.1 (a) Simple equilibrium energy band diagram of the ITO–SiO_2–p-Si system. E_{gos}, E_{gi} and E_{gs} denote the band gap of ITO, SiO_2, and Si respectively. ϕ_{osi} is the oxide semiconductor-to-insulator barrier height and is related to work function of ITO. ϕ_B is the barrier height which governs V_{oc} of the device. Adapted from Shewchun et al. (1978) (Ref. 101). (b) Schematic energy band diagram of the minority carrier ITO–SiO_2–p-Si system without illumination. The device is biased positively by a voltage V_A with respect to the top ITO layer. J_{CT} and J_{VT} are the current flows from the conduction and valence bands of the Si to the ITO layer. J_{ST} is a flow from the surface states to the ITO and is supplied by J_{CI} and J_{VI}. Also shown in the diagram are the electron and the hole Fermi levels in the semiconductors. Adapted from Shewchun et al. (1978) (Ref. 101). (c) Energy band diagram of the ITO–SiO_2–p-Si system near the maximum power conditions under illumination. Adapted from Shewchun et al. (1978) (Ref. 101).

semiconductor can be pinned effectively to their respective minority carrier bands over a limited bias range (101), as shown in the Fig. 3.2.1(b) when the bias voltage is increased, the semiconductor current increases to a value which can be supported by the tunneling process, giving the tunnel limited region of operation. In the beginning of this region, the electron quasi-Fermi level becomes unpinned with respect to the conduction band edge. Figure 3.2.1(c) shows a diode in a photovoltaic mode.

The variation of the hole and electron (J_{CT} and J_{VT}) currents with band gap of the oxide semiconductor has been investigated. For the surface-state free case, the tunnel current for holes (J_{VT}) vanishes as the energy gap gets larger. If the surface states are present, considerable tunnel current can be transported across the interfacial layer via surface states.

The ITO–SiO_2–pSi operates as a good device in the 10–30 Å range. Below 10 Å the Schottky diode behavior sets in while the upper limit is the value at which the ITO–Si diode would revert to the equilibrium mode of operation. In the nonequilibrium mode, the dominant current flow near the zero bias may be between the majority carrier bands, surface states, or the minority carrier bands. It has been shown (95) that the minority carrier diodes can be formed on p-type silicon by selecting materials with low oxide–semiconductor-to-insulator work functions. For interface thickness up to 16 Å the $I-V$ characteristics converge to the ideal Shockley diode equation up to forward bias voltages where the semiconductor current exceeds the tunnel current.

The barrier height of the device is given by (95):

$$\phi_b = E_{gs} + \phi_{si} - \phi_{\text{osi}} + \Delta - V_i + q\frac{(Q_s + Q_i)}{\varepsilon_i}d \qquad (3.2.1)$$

where V_i is the potential drop across the insulator, Q_s is the charge in the surface states, and Q_i is the fixed charge in the insulator. $\phi_{si} - \phi_{\text{osi}}$ is the discontinuity in the conduction band and should be negligible in order to get maximum open circuit voltage. In order for an SIS structure to have the maximum efficiency, the electron affinity of the oxide semiconductor should be equal to the p-type base semiconductor and for an n-type base semiconductor, the electron affinity should be equal to the sum of the band gap and the electron affinity of the n-type semiconductor (95). The oxide semiconductor should have large band gap in the range 3–4 eV, a condition which is met by SnO_2, ITO, and Cd_2SnO_4.

3.3 ITO-SnO_2–Si Solar Cells

Kajiyama and Furukawa (125) reported a SnO_2/Si heterostructure solar cell which was 0.1% efficient and was the first observation on a solar cell fabricated on silicon with a transparent oxide semiconductor, which was a subject matter of many subsequent investigations. Nishino and Hamakawa (126, 127) reported the short-circuit current and spectral response of the SnO_2/Si heterojunction solar cell. An energy-band diagram for the device was proposed. Photoelectric effects in In_2O_3–p-Si diodes have been reported by Matsunami et al. (128). Measurements of photoelectric affects were carried out on In_2O_3–p-Si as well as In_2O_3–n-Si. However, good rectification and large photoelectric effects were obtained only on In_2O_3–p-Si diodes. A simple fabrication procedure for SnO_2–Si solar cells was reported by Kato et al. (129). A silicon wafer substrate (*n-type*) of resistivity 0.7–1 ohm-cm was chemically etched in a solution of $HNO_3 : HF : CH_3COOH =$ 5:3:1 and then rinsed with deionized water and dried. Ohmic contact to the substrate was made by alloying an Au–Sb sheet on one side of the wafer. SnO_2 film was deposited on the other side of the wafer by spraying a solution of $SnCl_4$ in HCl at about 300°C. The spectral response of the short-circuit current was studied. The output current at long wavelength was found to be similar to Si p–n junction diodes, but the output current at short wavelengths was found to be larger than in silicon p–n junctions. This is because of the thin and transparent SnO_2 layer which allows the incident beam to pass through it almost unattenuated. A short circuit current of 0.285 mA/mm^2 and an open circuit voltage of 450 mV at 80,000 lux illumination have been reported for this device.

An efficiency close to 10% has been reported by Anderson (106, 130) for SnO_2/Si cells. He postulated a MIS type model to explain his results. The thin insulating layer in the SnO_2–Si device increases V_{oc}, the open circuit voltage, by decreasing J_0, the reverse saturation current. The dark current is due to electrons injected over the Si barrier through the oxide. In SnO_2–SiO_2–n-Si structures, under illumination, holes are trapped in the potential notch in the valence band, where they recombine either with electrons from SnO_2 and produce photocurrent, or with electrons from the Si conduction band via interface states and suppress the photocurrent.

ITO/Si heterojunction solar cells have been fabricated by RF sputtering ITO on n-Si by Mizrah and Adler (103, 104). This ITO deposition process yields good quality films with high transmission in solar region and low sheet resistivities (131, 132). The fabricated cells have larger response in the blue and UV region as compared to diffused silicon devices (133). The cells were fabricated by depositing a 4000 Å layer on to n/n^+ and p/p^+ Si epitaxial wafers in pure argon gas at a pressure of 17 m-torr at a power of 35 W. At an illumination level of 82 mW/cm^2, the efficiency was 1% and at 48 mW/cm^2, the efficiency was 1.5%. The photocurrent in ITO/n-Si has been attributed (103) to photoexcited electron–hole pairs in Si and the photo excited electrons from the conduction band of ITO through the SiO_x layer into the conduction band of Si. The photoexcited holes in the Si can contribute to the photocurrent by recombining with electrons from ITO conduction band via interface states. However, the recombination of the hole with photoexcited electrons in the Si conduction band would lead to the suppression of the photocurrent. The photoexcited electrons in Si are accelerated away from the interface and contribute to photocurrent. The low efficiencies of the cells in Mizrah and Adler's work (103) have been attributed to photocurrent suppression as first discussed by Anderson (106). A thin SiO_x layer leads to an enhanced open-circuit voltage as higher fields are required by the electron to tunnel through the SIO_x layer into ITO. The Fermi level in the ITO and Si will have to be separated by a larger amount in order that the forward current will balance the photocurrent, leading to an open-circuit photovoltage. A thick SiO_x layer does not allow holes in Si to tunnel through it and recombine with electrons in the ITO conduction band. In this situation the holes accumulate near the interface and recombine with photoexcited electrons, thus reducing the photo current. The larger is the intensity of illumination, the larger will be the number of holes trapped, and hence larger will be the rate of recombination, thus enhancing the photocurrent suppression.

Mizrah and Adler (104) have suggested the possibility of using amorphous silicon with an ITO layer. Such devices were found to exhibit efficiencies of 0.01% for the 100 mW/cm^2 white light. However, because of the large band gap of ITO, the devices show high photoresponse in the blue–UV region and the efficiencies are of the order of 0.5% for a light of 4000 Å wavelength.

DuBow et al. (91) reported ITO–*p*-Si heterojunction diodes of 12% efficiency produced by ion-beam technique. Figure 3.3.1 gives the fabrication diagram of an efficient ITO/*p*-Si solar cell. The cleaned wafer was in-situ etched in a vacuum system and the ITO was then sputter deposited by an ion-beam sputtering technique which allows the low-energy deposition of ITO on a substrate at a low background pressure. Substrate temperatures of 350–400°C were used. The wafer is metallized and contacts made. The devices were fabricated on 2 in. silicon wafers with seven solar cells per wafer. The process is also suitable for polycrystalline silicon. The best solar cells produced by this technique gave 13.2% efficiency under AMI sunlight. At a tin oxide concentration of 10% DuBow et al. obtained an open-circuit voltage of 0.51 V, short-circuit current of 32 mA/cm^2 and a fill factor of 0.70 for light close to sun's spectral distribution. The cells were fabricated on a 2 ohm-cm *p*-Si wafer on which a 4000 Å ITO film was deposited using an ion-beam technique. Oxide films were strongly adherent on silicon and glass and were found to be extremely abrasion resistant. The 90% In_2O_3 devices were reported to have good stability (91) with time. The chemical free energies of formation for In_2O_3, SnO_2, and SiO_2 are − 207, − 124,

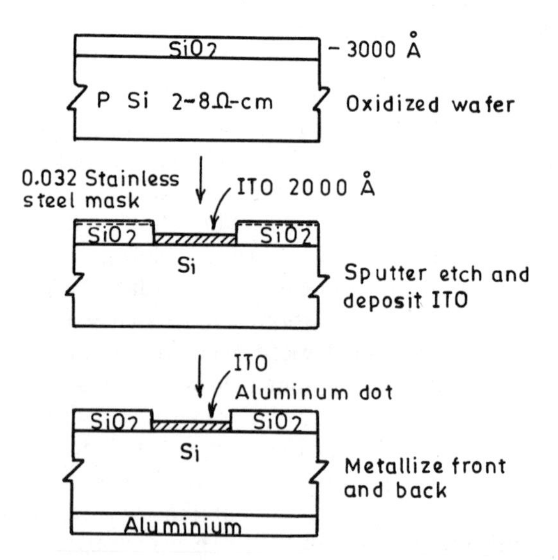

Fig. 3.3.1 Fabrication diagram of a ITO/*p*-Si solar cell. Adapted from DuBow et al. (1978) (Ref. 91a).

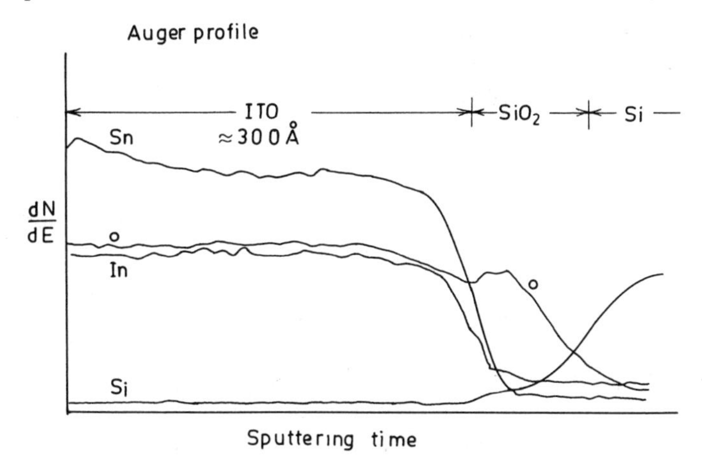

Fig. 3.3.2 Experimental Auger depth profile of a solar cell. Adapted from DuBow et al. (1978) (Ref. 91a).

and − 193 kcal/mole, respectively (134), and therefore In_2O_3 is expected to be stable on a silicon substrate whereas SnO_2 may give up oxygen to silicon (91). DuBow et al. (91) explained their observations on a *p*-type Schottky barrier model for the photovoltaic cell proposed by Anderson (106). DuBow et al. (91a) also performed the Auger depth profile analysis. A typical profile is shown in Fig. 3.3.2. There is an evidence of interfacial region between ITO and the silicon. It was concluded that for oxide thicknesses greater than 20 Å, the interface was SiO_2. This is the first experimental evidence of SiO_2 layer in ITO/Si solar cell.

Manifacier and Szepessy (107) reported In_2O_3:Sn–*n*-type Si heterojunction solar cells where the transparent and conductive In_2O_3:Sn layer was made by simple and cheap spray pyrolytic process. Under AMI simulated sunlight an open-circuit voltage, $V_{oc} = 500$ mV, short-circuit current, $I_{sc} = 32$ mA/cm^2, and fill factor 0.6 to 0.65 with conversion efficiency of 10% have been reported.

The solar cells were fabricated using *n*-type Si substrates with resistivities between 1–10 ohm-cm. An aerosol stream containing an alcoholic solution of $SnCl_4 \cdot 5H_2O$ and $InCl_3$ was sprayed through a preheating furnace. From the value of $n \approx 2$–2.7 deduced for the diode quality factor, they obtained low value of the Richardson's constant, and the observed decrease in the efficiency yield versus illumination suggest a MIS model for these structures (98). The abrupt heterojunction model developed by Lai et al. (108) for the In_2O_3 evaporated silicon (*p*-type) structures is in contradiction with the results of Manifacier and Szepessy (107) as poor open-circuit voltage was obtained when *p*-Si was used.

Ghosh et al. (135) fabricated highly efficient SnO_2/Si devices. The dark I–V characteristics were attributed to a combination of thermionic and diffusion processes. An MIS model was used to explain the spectral response curve, the short-circuit photocurrent, the open-circuit photovoltage, and the light and dark *I–V* characteristics. The silicon used was *n*-type, usually 1–3 ohm-cm resistivity and about 10–11 mil thick. SnO_2 of 500–700 Å thickness was deposited by electron-beam evaporation.

Feng et al. (136) have recently fabricated cells with efficiencies more than 12% of SnO_2–Si by electron-beam deposition or spraying. They report a barrier height of 0.8 eV from an analysis of the reverse saturation current. From the UV photoemission spectroscopy studies, the work function of SnO_2 was estimated to be 4.85 eV and was found to be strongly dependent on the stoichiometry and oxygen content.

Wang and Legge (92) investigated photovoltaic effect in *n*-SnO_2–*n*-Ge and *n*-SnO_2–*p*-Ge heterojunctions produced by CVD method. In (*n–n*) SnO_2 – Ge they observed J_{sc} of 38 mA/cm^2, V_{oc} of 77 mV, and a fill factor of 0.30. The polarity of V_{oc} and J_{sc} with respect to base material was positive. For (*n–p*) SnO_2–Ge heterojunctions, no photovoltaic effect was observed and the *I–V* characteristic was ohmic in low bias range. The ohmic behavior was attributed to an accumulation layer between SnO_2 and Ge (149).

3.4 *n*-ITO/*p*-CdTe Heterojunctions

Bube (93) studied photovoltaic effect in *n*-ITO/*p*-CdTe heterojunctions produced by sputtering ITO onto *p*-CdTe single crystals. A V_{oc} of 0.82 V, J_{sc} = 145 mA/cm^2, fill factor 0.55, and a solar efficiency of 8% have been obtained.

3.5 *n*-ITO/*p*-InP Solar Cells

High-efficiency *n*-ITO/*p*-InP solar cells with amorphous and crystalline ITO have been fabricated by Sree Harsha et al. (121) and Bachmann et al. (112). In view of the large mismatch (124) of crystal structures, lattice parameters, and thermal expansion coefficients (150) of ITO and InP, they are not expected to form a good device and therefore the observed high efficiencies are surprising. A plausible explanation of the observed results has been offered by Singh and Shewchun (124) on SIS model. The 14.4% efficiency observed in such devices has been interpreted as due to the possible presence of a thin P_2O_5 layer at the interface. It has been shown that (124) in the presence of thin insulating layer, an efficiency of 25.6% is possible in such a structure. Amorphous ITO cells performed better than crystalline ones, as the lower substrate temperatures which allow the amorphous ITO to form are better for the formation of the insulating layer of P_2O_5. Wilmsen and Kee (151) did Auger analysis of In_2O_3/InP interface and observed In–O bonding on ITO side and In–P bonding on the InP side and the transition region consisted of *P*-O bonding. The possibility that the transition region consists of a ternary compound is also not ruled out (124). However, further work is necessary to establish the presence of P_2O_5 layer.

3.6 ITO, SnO$_2$/GaAs Solar Cells

Wang and Legge (92) have investigated the (*n–n*) SnO_2 – GaAs and (*n–p*) SnO_2 – GaAs heterojunctions produced by chemical vapor deposition. In (*n–n*) GaAs heterojunction under AMO condition a J_{sc} of 11.4 mA/cm^2, V_{oc} of 0.33 V and a fill factor (FF) of 43% for cells of series resistance 6 ohms were obtained. Under illumination, the polarity of V_{oc} and I_{sc} was positive for the SnO_2 side and negative for GaAs side. For (*n–p*) SnO_2–GaAs cells, the photovoltaic effect was small. Hsu and Wang (113) reported the successful fabrication of In_2O_3–GaAs heterojunctions. The observed polarity of V_{oc} and I_{sc} was found to be consistent with the band bending of a simple In_2O_3/semiconductor band diagram. An electron affinity of 4.45 eV for In_2O_3 is reported. For (*n–p*) In_2O_3–GaAs heterojunctions, J_{sc}, V_{oc}, and FF were found to be 6.17 mA/cm^2, 0.084 V, and 0.28, respectively, under AMO illumination. The (*n–n*) In_2O_3/GaAs showed good photovoltaic properties. These cells fit in the abrupt heterojunction model. Although the nature of the interface layer is not known, it is thought to be Ga_2O_3, As_2O_3, or a mixture of the two (95). The efficiency of *n*-ITO–Ga_2O_3–pGaAs as a function of insulator thickness has been computed theoretically (Fig. 3.6.1). An efficiency of 20% can be achieved, although no efficient cells have so far been made. Oxide semiconductor solar cells consisting of indium tin oxide and *n*-GaAs were fabricated by Agnihorti et al. (113a). ITO was deposited by a pyrolytic decomposition method onto *n*-GaAs substrates. Back ohmic contact with GaAs was established with a Au–12% Ge alloy.

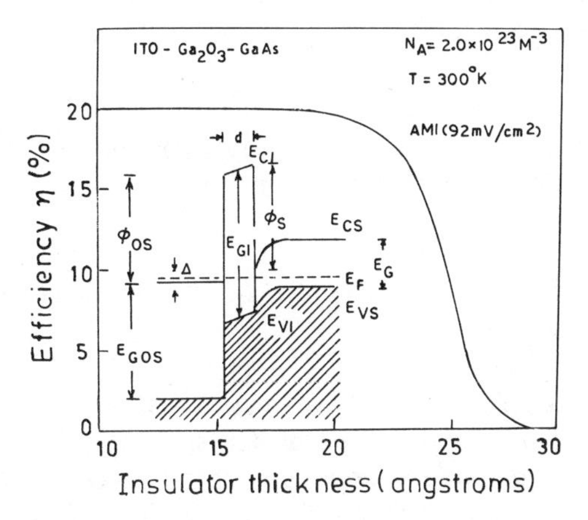

Fig. 3.6.1 Computed effect of insulator thickness of an ITO–Ga_2O_3–*p*-GaAs tunnel diode under AMI illumination. About 20% reflection loss from ITO has been taken. Adapted from Shewchun et al. (1978) (Ref. 95).

At 100 mw/cm^2 we have observed an open-circuit voltage of 0.918 V, short-circuit current of 21.2 mA/cm^2, a fill factor of 0.52, and conversion efficiency of 10.7%. The results are in disagreement with general SIS model proposed by Shewchun et al. (95) which predicts that n-GaAs is not likely to form a good device with ITO. The electron affinity of ITO prepared by pyrolytic decomposition is perhaps more than the value proposed by Hsu and Wang (113) and is the likely cause of higher efficiencies observed in this investigation. The possibility that sprayed films will have higher electron affinity has also been suggested by Shewchun et al. (95). The electron affinity of oxide semiconductor is given by

$$\chi_{os} = \chi_{os}^{(e)} + \chi_{os}^{(p)} \tag{3.6.1}$$

where $\chi_{os}^{(e)}$ is the contribution from the electrostatic part and is the work against any permanent dipole layers at the oxide semiconductor surface. $\chi_{os}^{(p)}$ is the contribution from the polarization part and is the work against the image force. The contribution $\chi_{os}^{(e)}$ depends upon the surface of the oxide semiconductor and will therefore be different in films prepared by different methods.

3.7 $CuInSe_2$/ITO Solar Cells

Kazmerski and Sheldon (114) have recently reported the successful fabrication of ITO/$CuInSe_2$ heterojunction solar cells with AMI efficiency of 8.5% when the base semiconductor is a single crystal. When the base is a thin film, the conversion efficiency is 2.08%. $CuInSe_2$, with a band gap of 1.02 eV, is a suitable material for photovoltaic heterojunction formation with ITO (114). The ITO films in this work were produced by electron-beam deposition. Better device performance has been observed for devices with ITO deposited at a substrate temperature of 180°C. A study of the degradation in these devices has been performed by measuring J_{sc} when the device was held at 200°C. J_{sc} falls off rapidly after 10 hr. An interface oxide layer (SeO_x or Cu_xO) is the likely cause of this degradation. Further work is needed to understand the degradation phenomena.

3.8 Role of Electron Affinity of Oxide Semiconductors as Used in Solar Cells

Thomson and Anderson (152) have reported measurements on ITO/Si heterojunction solar cells produced by RF sputtering in an argon atmosphere. The conduction band discontinuity at the interface, ΔE_c was found to be of the order of 0.45 eV for ITO with 9 mole % SnO_2. The electron affinity of ITO containing 9 mole % SnO_2 is therefore about 0.45 eV greater than that for Si and is about 4.5 eV. It has been concluded that due to the magnitude of electron affinity, the built in voltages in both ITO/p-Si and ITO/n-Si cells will be insufficient for solar energy conversion and ITO is therefore not a suitable material for making solar cells. It appears therefore that the electrical and photovoltaic characteristics of RF sputtered ITO/Si heterojunctions are sensitive to sputtering parameters. However, Singh and Shewchun (153) have pointed out that in the presence of an interfacial layer, the conclusions and the analysis given by Thomson and Anderson will not be valid. The electron affinity of ITO without interface is suggested to be 4.3 eV (153). From the analysis of the dark $I-V$ curves, which consisted of two regions, one with $n_1 \approx 1.3$ and the other with $n_2 \approx 2.0$, i.e., the thermionic emission and depletion recombinations, the barrier height of the device was determined to be 0.85. The electron affinity of oxide semiconductor in the absence of any interfacial layer and surface states at the interface is given by (153):

$$\chi_{os} = \chi_s + \Delta + E_{gs} - \phi_b \tag{3.8.1}$$

where $\phi_b = V_D + \delta$ (Fig. 3.8.1). Taking $\Delta = 0$, $\chi_s = 4.05$, and $E_{gs} = 1.1$ eV we find that $\chi_{os} = 4.3$ eV. This value of electron affinity is nearly the correct value required for minimum discontinuity. The value of electron affinity

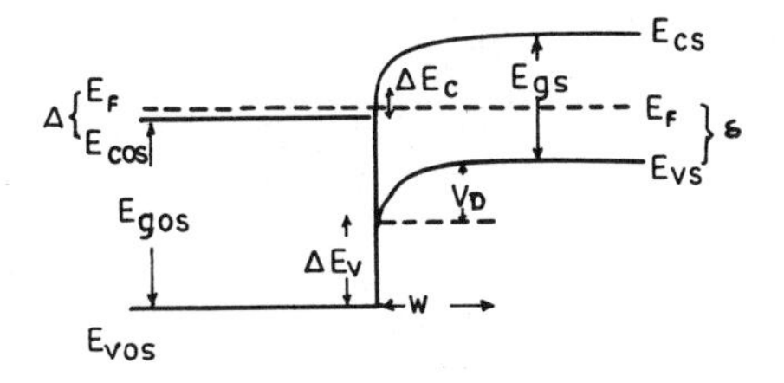

Fig. 3.8.1 Simple equilibrium energy band diagram for n-oxide semiconductor/p-base–semiconductor solar cell, neglecting the presence of an interfacial layer. Adapted from Singh and Shewchun (1978) (Ref. 153).

Table 3.8.1

Semi-conductor	Type of Crystal Structure	Lattice Constants (Å) *a*	*b*	*c*	Linear Coefficient of Thermal Expansion (10^{-6}/°C)
Si	Diamond	5.431	—	—	2.33
Ge	Diamond	5.657	—	—	5.8
InP	Zinc blende	5.869	—	—	4.5
GaAs	Zinc blende	5.653	—	—	5.8
CdTe	Zinc blende	6.477	—	—	5.9
$CuInSe_2$	Chalcopyrite	5.782	—	11.62	—
In_2O_3	Cubic	10.118	—	—	10.2
SnO_2	Tetragonal	4.737	—	3.185	4.0
ITO	Cubic	10.118	—	—	10.2
ZnO	Hexagonal	3.249	—	5.205	—

Source: Adapted from Singh and Shewchun (153).

Table 3.8.2 Data on Transparent Oxide Semiconductor/Semiconductor Solar Cells

Cell Base/Front	Crystalline State S: Single Crystal A: Amorphous P: Polycrystalline	E_{gb}/E_{gfr} (eV)	V_{oc} (V)	J_{sc} (mA/cm^2)	FF	η %	Illum. AMX or (mW/cm^2)	Antireflection Coating	Reference	Remarks
p-Si/n-In_2O_3	S(111)/P	1.11*i*/3.6*i*	0.16	0.73			1000^+ lux^+	no	128, 137	
n-Si/n-In_2O_3	S(100)/P	1.11*i*/3.6*i*	0.23	0.68	0.53		1000^+ lux^+		137, 138	
p-Si/ITO	S	1.11*i*/3.6*i*	0.51	32	0.70	12	92 AM	no	91, 139	ITO was prepared by ion-beam technique
n-Si/ITO	S	1.11/3.6*i*	0.50	32	0.65	11	1.2		103, 106, 107	
p-Si/ITO	P	1.11*i*/3.6*i*	0.28	25	0.23	1.6	100^+		140	
$i(n)$Si/ITO	A	1.55/3.6*i*	0.43	10	0.28	1.2	100^+	no	141	*a*-Si by glow discharge
Si/ITO	A	1.55/3.6*i*				0.01– 0.02	100^+		104	*a*-Si sputtered
p-Si/n-SnO_2	S	1.11*i*/3.5*i*				1.15	AMI^+		116, 142	
n-Si/n-SnO_2	S	1.11*i*/3.5*i*	0.521	29	0.64	9.9	AMI^+		116, 129, 142, 143	
n-Si/n-SnO_2	P	1.11*i*/3.5*i*	0.463	26	0.60	7.2	AMI^+		116	

p-CdTe/n-ITO	S		0.82	14.5	0.55	8	85		93	Sputtered ITO
p-GaAs/n-In_2O_3		1.43d/3.6i	0.084	6.2	0.40		AMO^+	no	137	
n-GaAs/n-In_2O_3		1.43d/3.6i	0.082	1.2			AMO^+	no	137	
p-GaAs/n-SnO_2		1.43d/3.5i	0.006	0.019	0.25				105	
n-GaAs/n-SnO_2		1.43d/3.5i	0.33	11.4	0.43	1.2	AMO^+		105	
n-InP/n-In_2O_3		1.43d/3.6i	0.018	0.6	0.35		AMO^+	no	137	
p-InP/a-ITO	S	1.43d/3.6i	0.76	21.55	0.65	14.4	AM2	MgF_2	121	
$Ga_{1-x}Se_x/SnO_2$	A	2.50/3.5i	0.65	0.002		3×10^{-3}	10^+		144	
GeSe/Se/n-SnO_2	A	/3.51				10^{-3}			145	
$Ge_{1-x}Se_x/SnO_2$	A	/3.51				3×10^{-5} ($x = 0.5$)	7.8^+	SnO_2 glass	146	
p-$In_{1-x}Se_x$/n-SnO_2	A	1.74/3.5i	0.10	0.01		0.03	7.8^+	SnO_2/ glass	145, 147	
p-Si/ITO	P	1.11i/3.6i	0.48	29	0.65	9	AM2		148	neutralized ion-beam sputtering of ITO.
p-GaAs/ITO	S	1.43d				5	AM2		112	ITO sputtered
$CuInSe_2$/ITO	S	1.01d/3.6i		30		8.9	100^+		114	ITO: 9% SnO_2 + 91% In_2O_3
$CuInSe_2$/ITO	P	1.01d/3.6i				3–5.4	100^+		114	ITO by vacuum evaporation
n-Si/SnO_2	S	1.11i/3.5i	0.51	33–39	0.40–0.55	10–12	AM1		136	SnO_2 electron-beam evaporated and spray
n-Ge/SnO_2	S		0.77	38	0.30		AM1		92	SnO_2 formed by CVD

will therefore depend upon the value of Δ. Heat treatment of ITO films change the value of Δ to the extent of 0.4 eV (154).

In the presence of surface states the electron affinity is given by (95)

$$\chi_{os} = \chi_s + \Delta + E_{gs} - \phi_b - V_i + q\left[\frac{Q_s + Q_i}{\varepsilon_i}\right]d \tag{3.8.2}$$

where V_i is the potential drop across the insulator, Q_s is the charge in the surface states, and Q_i is the fixed charge in the insulator. This value, in the presence of interfacial layer is calculated to be 4.2 eV (153).

Table (3.8.1) (modified from Ref. 153) gives the data on crystal structure, lattice parameters, and thermal expansion coefficients of oxide semiconductors and some base semiconductors. Any combination of an oxide semiconductor and base semiconductor has appreciable lattice mismatch, and is therefore not likely to form a good device. The absence of interfacial layer and surface states at the interface can therefore lead to uncertainties in the value of electron affinity of oxide semiconductors and this is likely to be the cause of the disagreement of the Thomson and Anderson (152) results with earlier observations reporting high efficiency of ITO/Si cells (91). The data on the transparent oxide semiconductor solar cells is tabulated in Table 3.8.2. This is partly adapted from Bucher (155) and updated to give the latest results.

3.9 Degradation in Transparent Oxide Semiconductor Solar Cells

Notable contribution in this area come from Anderson and his group (117, 118) who have investigated degradation, accelerated life tests, and intensity effects in SnO_2/Si heterojunction solar cells. The reduction of SnO_2 by Si produces a thin insulating layer of SiO_2 (116) which is responsible for increasing the series resistance of the cells. A slow degradation of SnO_2–Si cells has been observed at elevated temperatures and is believed to be the result of growth of the insulating layer with time. The use of polycrystalline silicon produces thicker insulating layers. This layer is responsible for suppressing the photocurrent by trapping and recombination of photogenerated carriers in the base semiconductor via interface states thus reducing the collection efficiency (106). This excludes the possibility of using these cells with concentrated sunlight. The saturated photocurrent in SnO_2–*n*-Si cells was found to be linear with illumination intensity up to 30 suns (117).

3.10 Effect of Surface States and Surface Charge

Although the origin of surface states and surface charge in an Si–SiO_2 interface is not very well understood, these effects are known to strongly influence the characteristics of oxide semiconductor solar cells (100). Surface states are charge storage and recombination-generation centers and provide additional tunneling paths. The presence of a surface charge in the insulator region causes a change in the oxide semiconductor insulator work function. Calculations for insulator thicknesses of 14 Å and less have shown that smaller surface states and surface charges lead to larger open-circuit voltages (100). This is because of the reduction in the barrier height in the presence of surface states and surface charge. The surface states and surface charge are thought to originate in the device processing conditions and this needs further investigation.

3.11 Future Possibilities of Oxide Semiconductor Solar Cells

SIS solar cells are capable of high conversion efficiencies. Due to low-temperature processing steps, they open up the possibility of using polycrystalline silicon in future technology. Fabrication of integrated arrays can cut down drastically the cost of silicon solar cells. The future work should aim at achieving 10% efficiencies with oxide-semiconductor–insulator–polycrystalline-silicon solar cells. Techniques for growing stoichiometric ultrathin oxides have to be developed to achieve highest conversion efficiencies. Direct evidence for the existence of a SiO_x layer has recently been obtained by Auger electron spectroscopy (91a, 156). An accurate control of the interfacial layer and stoichiometry has been achieved by DuBow et al. (91a, 156) who reported J_{sc} of 36 mA/cm^2, V_{oc} of 0.50 V, FF of 0.74, and conversion efficiencies of 13.2%. For large-scale terrestrial applications as a solar energy converter, the cost of the device will be determined by the cost and availability of indium. The question of indium availability has been examined (91a) and it is anticipated that indium supply will not limit the large-scale terrestrial applications of ITO/Si solar cells.

4

Characterization of Selective Surfaces

In this chapter, the basic concepts of absorptance, emittance, and reflectance for opaque, specular, and diffuse surfaces are discussed. Solar absorptance and thermal emittance are two basic parameters for characterizing the selective nature of various coatings. A detailed survey is made about the measurements of solar absorptance and thermal emittance by direct and indirect techniques. In indirect determination, the reflectance of a surface is measured in different ranges of the spectrum by the integrated sphere investigated by Gier and Dunkle and many other researchers. Here the measurements are quick but are dependent on the properties of reference sample. Calorimetric methods for the absolute measurement of emittance are also discussed. A brief discussion about the solar calorimeter is given for the direct determination of α/ε. Commercial portable instruments for the determination of α and ε are also described.

4.1 Absorptance and Emittance

The absorptance α of a plane surface is the fraction of incident radiation which is absorbed by the surface. If the surface is opaque to the radiation then absorptance and reflectance sum is unity. Both absorptance and reflectance are functions of the wavelength of radiation and the angle of incidence. Therefore the monochromatic directional absorptance is defined as the fraction of the incident radiation of wavelength λ from the direction μ, ϕ (μ is consine of polar angle and ϕ is the azimuthal angle that is absorbed by plane surface). The radiant energy which reaches elemental area dA from the (θ, ϕ) direction is shown in Fig. 4.1.1. The expression for absorptance is given by

$$\alpha_\lambda(\mu, \phi) = \frac{J_{\lambda,a}(\mu, \phi)}{J_{\lambda,i}(\mu, \phi)} \tag{4.1.1}$$

where a and i are subscripts used for absorbed and incident radiations and $J_\lambda(\mu, \phi)$ is spectral irradiance of the radiation.

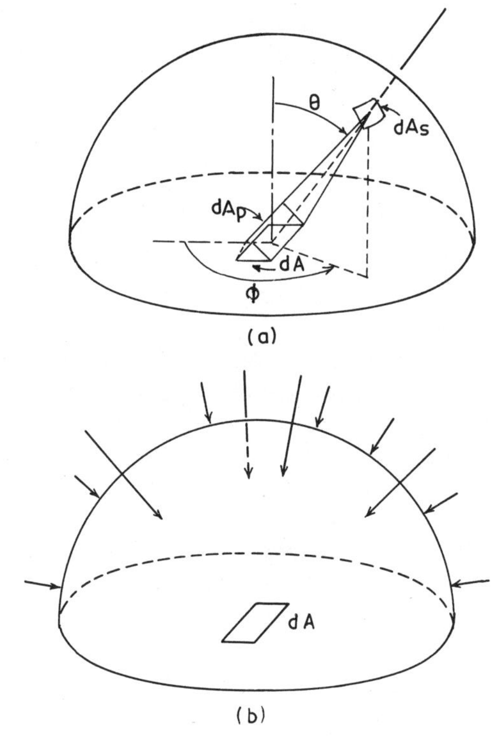

Fig. 4.1.1 Geometrical representation of absorbed radiant energy. (a) Directional absorptivity; (b) hemispherical absorptivity.

The directional absorptance is defined as the fraction of all the radiations absorbed by the surface from μ, ϕ direction and is defined by the following expression:

$$\alpha(\mu,\phi) = \frac{\int_0^\infty \alpha_\lambda(\mu,\phi) J_{\lambda,i}(\mu,\phi)\, d\lambda}{\int_0^\infty J_{\lambda,i}(\mu,\phi)\, d\lambda} \tag{4.1.2}$$

or

$$\alpha(\mu,\phi) = \frac{\int_0^\infty \alpha_\lambda(\mu,\phi) J_{\lambda,i}(\mu,\phi)\, d\lambda}{J_i(\mu,\phi)} \tag{4.1.3}$$

The monochromatic hemispherical absorptance is obtained by integrating the above expression over the enclosing hemisphere and is given by the following expression:

$$\alpha_\lambda = \frac{\int_0^{2\pi}\int_0^1 \alpha_\lambda(\mu,\phi) J_{\lambda,i}(\mu,\phi)\mu\, d\mu\, d\phi}{\int_0^{2\pi}\int_0^1 J_{\lambda,i}(\mu,\phi)\mu\, d\mu\, d\theta} \tag{4.1.4}$$

Similarly the hemispherical absorptance can be obtained simply by integrating the above expression over all wavelengths.

$$\alpha = \frac{\int_0^\infty \int_0^{2\pi} \int_0^1 \alpha_\lambda(\mu, \phi) J_{\lambda,i}(\mu, \phi) \mu \, d\mu \, d\phi \, d\lambda}{\int_0^\infty \int_0^{2\pi} \int_0^1 J_{\lambda,i}(\mu, \phi) \mu \, d\mu \, d\phi \, d\lambda} \tag{4.1.5}$$

If the surface is not plane, it will have an effective solar absorptance given by the ratio of the solar power absorbed by the surface to the solar power incident on the surface.

The emittance is described as the ability of a surface to radiate thermal power. It is the ratio of the radiancy of the surface to the radiancy of the blackbody at the same temperature and under the same conditions. The geometrical representation of directional and hemispherical emittance is shown in Fig. 4.1.2. Spectral emittance is the ratio of the spectral radiancy (or monochromatic radiancy at a given wavelength) from a surface to that of a blackbody at the same temperature and is given by the expression:

$$\varepsilon_\lambda = \frac{J_\lambda}{J_{b\lambda}} \tag{4.1.6}$$

The spectral directional emittance of a surface is defined as the ratio of monochromatic intensity by a surface in a particular direction to the mono-

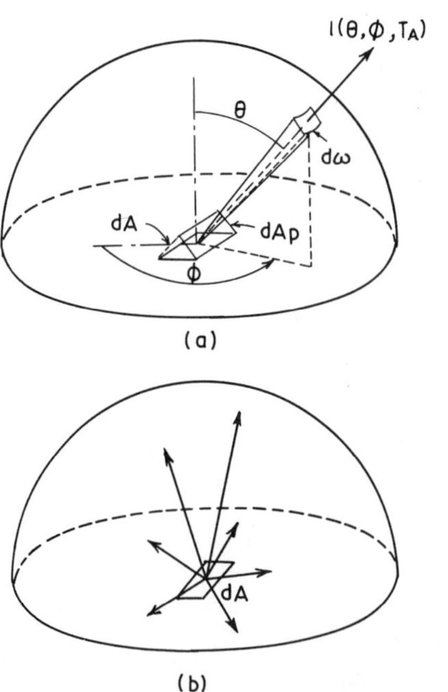

Fig. 4.1.2 Geometrical representation of directional and hemispherical radiation properties. (a) Directional emissivity; (b) hemispherical emissivity.

chromatic intensity emitted by a blackbody at the same temperature and can be expressed as

$$\varepsilon_\lambda(\mu,\phi)=\frac{J_\lambda(\mu,\phi)}{J_{b\lambda}(\mu,\phi)} \tag{4.1.7}$$

The total emittance $\varepsilon_t(\mu,\phi)$ is defined as the ratio of the radiancy of the surface to the radiancy of the blackbody at the same temperature and includes all wavelengths from zero to infinity:

$$\varepsilon_t(\mu,\phi)=\frac{\int_0^\infty \varepsilon_\lambda(\mu,\phi)J_{b\lambda}\,d\lambda}{\int_0^\infty J_{b\lambda}\,d\lambda}$$

$$=\frac{1}{J_b}\int_0^\infty \varepsilon_\lambda(\mu,\phi)J_{b\lambda}\,d\lambda \tag{4.1.8}$$

Hemispherical emittance $\varepsilon_{h\lambda}$ refers to the emission in all possible directions, i.e., 2π steradians, and is given by

$$\varepsilon_{h\lambda}=\frac{\int_0^{2\pi}\int_0^1 \varepsilon_\lambda(\mu,\phi)J_{b,\lambda}\mu\,d\mu\,d\phi}{\int_0^{2\pi}\int_0^1 J_{b\lambda}\mu\,d\mu\,d\phi}$$

$$=\frac{1}{\pi}\int_0^{2\pi}\int_0^1 \varepsilon_\lambda(\mu,\phi)\mu\,d\mu\,d\phi \tag{4.1.9}$$

The total hemispherical emittance is given by

$$\varepsilon_h=\frac{\int_0^\infty\int_0^{2\pi}\int_0^1 \varepsilon_\lambda(\mu,\phi)J_{b,\lambda}\mu\,d\mu\,d\phi\,d\lambda}{\int_0^\infty\int_0^{2\pi}\int_0^1 J_{b,\lambda}\mu\,d\mu\,d\phi\,d\lambda} \tag{4.1.10}$$

The normal emittance ε_n (a special case of directional emittance) refers to the emittance normal to the surface (i.e., to the direction of the axial ray of a beam that is encompassed by a small solid angle) and is the ratio of the normal sterdiancy of a surface to that of blackbody at the same temperature.

The total hemispherical emittance, normal spectral emittance, and total normal emittance show different surface properties and also denote a specified method of measurement.

4.2 Reflectance

The problem of specifying the reflectance from a surface is more complex than the emittance and absorptance. The surface may be irradiated by radiations whose spectral intensity varies with the angle of incidence. Sometimes only spectral reflectance at the given angle of incidence may be sufficient, but in some instances the spatial as well as spectral distribution of the reflected radiation is required. When the incident radiation is in the form of narrow pencil (a small solid angle), two limiting distributions of the reflected radiation exist. The two limiting distributions are called "specular" and "diffuse." Figure 4.2.1 shows reflection of different kinds from a surface. In case of specular reflection the incident polar angle is equal to the reflected polar angle and the azimuthal angle differ by 180°. On the other hand in the case of diffuse reflectance, the diffuse reflections obliterate all directional characteristics of the incident radiation by distributing the radiation uniformly in all directions. The ratio of the reflected intensity in the θ_r, ϕ_r direction to that which is incident on a surface in the θ_i, ϕ_i direction at wavelength λ is termed the "bidirectional spectral reflectivity"; this may be defined as:

$$\rho_\lambda(\mu_r, \phi_r, \mu_i, \phi_i) = \frac{J_{\lambda,r}(\mu_r, \phi_r, \mu_i, \phi_i)}{J_{\lambda,i}(\mu_r, \phi_r, \mu_i, \phi_i)} \tag{4.2.1}$$

The bidirectional spectral reflectivity can be summed over the wavelength range to give the total reflectivity. It can be summed over all reflected solid angles, over all incident solid angles, and over both incident and reflected solid angles to give various combinations of directional and hemispherical reflectivities. Siegel and Howell (157) and Bevans and Edwards (158) present a comprehensive analysis of the problem.

Two types of hemispherical reflectance exist. The angular hemispherical reflectance is found when a narrow pencil of radiation is incident on a surface and all the reflected radiation is collected. The hemispherical angular reflectance results from collecting reflected radiation in a particular direction when the surface is irradiated from all directions. The spectral angular hemispherical reflectance is defined as the ratio of the monochromatic radiant energy reflected in all directions to the incident radiant flux within a small solid angle $\Delta\omega_i$. The incident energy $J_{\lambda,i}\mu_r\Delta\omega_i$ that is reflected in all

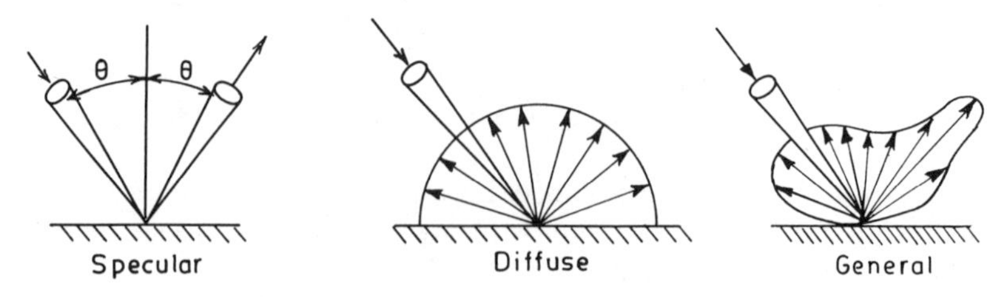

Fig. 4.2.1 Reflection from a surface.

directions can be calculated by using the reflection function:

$$q_{\lambda,r} = \frac{1}{\pi}\int_0^{2\pi}\int_0^1 \rho_\lambda(\mu_r, \phi_r, \mu_i, \phi_i) \times J_{\lambda,i}\mu_i\mu_r\Delta\omega_i d\mu_r d\phi_r \tag{4.2.2}$$

The spectral angular hemispherical reflectance can be expressed as

$$\rho_\lambda(\mu_i, \phi_i) = \frac{q_{\lambda,r}}{J_{\lambda,i}\mu_i\Delta\omega_i} = \frac{1}{\pi}\int_0^{2\pi}\int_0^1 \rho_\lambda(\mu_r, \phi_r, \mu_i, \phi_i)\mu_r d\mu_r d\phi_r \tag{4.2.3}$$

The angular hemispherical reflectance can be calculated by integrating the incident and reflected flux over all wavelengths.

The spectral hemispherical angular reflectance is defined as the ratio of the reflected monochromatic intensity in the direction (μ_r, ϕ_r) to the monochromatic energy from all directions divided by π.

The incident energy may be written in terms of incident intensity integrated over the hemisphere:

$$q_{\lambda,i} = \int_0^{2\pi}\int_0^1 J_{\lambda,i}\mu_i d\mu_i d\phi_i \tag{4.2.4}$$

and the spectral hemispherical angular reflectance is then

$$\rho_\lambda(\mu_r, \phi_r) = \frac{\pi J_{\lambda,r}(\mu_r, \phi_r)}{q_{\lambda,i}}$$

or

$$\rho_\lambda(\mu_r, \phi_r) = \frac{\int_0^{2\pi}\int_0^1 \rho_\lambda(\mu_r, \phi_r, \mu_i, \phi_i) J_{\lambda,i}\mu_i d\mu_i d\phi_i}{\int_0^{2\pi}\int_0^1 J_{\lambda,i}\mu_i d\mu_i d\phi_i} \tag{4.2.5}$$

The hemispherical angular emittance depends upon the angular distribution of the incident intensity. When the incident radiations are diffuse, the spectral hemispherical angular reflectance is identical to the spectral angular hemispherical reflectance. The total hemispherical angular reflectance and total angular hemispherical reflectance may be calculated by integration over all wavelengths. In case of angular hemispherical reflectance one gets:

$$\rho(\mu_i, \phi_i) = \frac{\int_0^\infty q_{\lambda,r} d\lambda}{\int_0^\infty J_{\lambda,i}\mu_i \Delta\omega_i d\lambda}$$

$$= \frac{1}{\pi J_i}\int_0^\infty\int_0^{2\pi}\int_0^1 \rho_\lambda(\mu_i, \phi_i, \mu_r, \phi_r) J_{\lambda,i}\mu_r d\mu_r d\phi_r d\lambda \tag{4.2.6}$$

When the surface element dA is irradiated from all directions and all the reflected radiation is collected, the reflectivity is characterized by the spectral hemispherical reflectance, defined as

$$\rho_\lambda = \frac{q_{\lambda,r}}{q_{\lambda,i}} \tag{4.2.7}$$

The reflected monochromatic energy $q_{\lambda,r}$ can be expressed as

$$q_{\lambda,r} = \int_0^{2\pi}\int_0^1 \left[\int_0^{2\pi}\int_0^1 \frac{\rho_\lambda}{\pi}(\mu_r, \phi_r, \mu_i, \phi_i) \times J_{\lambda,i}\mu_i d\mu_i d\phi_i\right]\mu_r d\mu_r d\phi_r \tag{4.2.8}$$

and the incident energy can be expressed as

$$q_{\lambda,i} = \int_0^{2\pi}\int_0^1 J_{\lambda,i} u_i du_i d\phi_i \tag{4.2.9}$$

Therefore the spectral hemispherical emittance can be expressed as

$$\rho_\lambda = \frac{\int_0^{2\pi}\int_0^1\left[\int_0^{2\pi}\int_0^1 \frac{\rho_\lambda}{\pi}(\mu_r, \phi_r, \mu_i, \phi_i) J_{\lambda,i}\mu_i \times d\mu_i d\phi_i\right]\mu_r \times d\mu_r d\phi_r}{\int_0^{2\pi}\int_0^1 J_{\lambda,i}\mu_i d\mu_i d\phi_i} \tag{4.2.10}$$

The hemispherical emittance may be expressed as

$$\rho = \frac{q_r}{q_i} = \frac{\int_0^\infty q_{\lambda,r} d\lambda}{\int_0^\infty q_{\lambda,i} d\lambda} \tag{4.2.11}$$

The hemispherical emittance depends on both the angular distribution and wavelength distribution of the incident radiation.

4.3 Relationship among Reflectance, Emittance, Absorptance, and Kirchhoff's Law

Let us consider a radiant energy interchange when a small body is placed in an isothermal enclosure or hohlraum and allowed to attain equilibrium. The isothermal enclosure is maintained at temperature T(in °K). In order to avoid violation of the second law of thermodynamics, if is found that spectral absorptance must be equal to spectral emittance. For an opaque body

$$\alpha_\lambda(\mu, \phi) = \varepsilon_\lambda(\mu, \phi) = 1 - \rho_\lambda(\mu, \phi) \tag{4.3.1}$$

By integration one gets that hemispherical characteristics also follow in same relationships:

$$\alpha_\lambda = \varepsilon_\lambda = 1 - \rho_\lambda \tag{4.3.2}$$

Thus the spectral emittance and spectral absorptance can both be calculated from the knowledge of angular hemispherical reflectance.

4.4 Measurement of Solar Absorptance and Thermal Emittance

Various techniques are used in solar energy and aerospace industries for measuring solar absorptance and thermal emittance of solar selective surfaces. The measured values are used to define the energy balance, and hence the temperature of the absorbing surface. To determine the values,

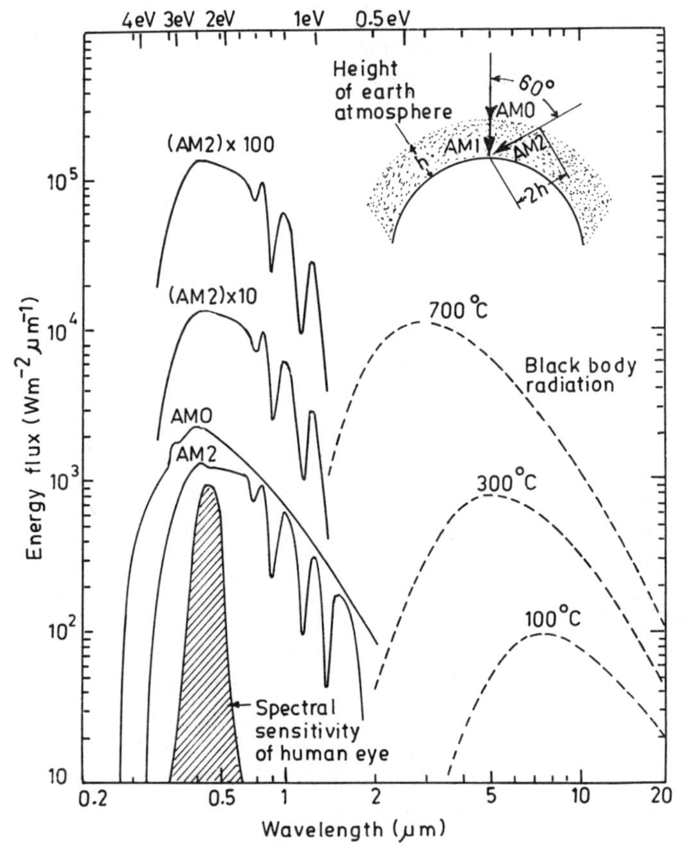

Fig. 4.4.1 Spectral distribution of sun's radiations at AMO and AM2 and blackbody's radiation emitted at 100 °C, 300 °C, and 700 °C.

measurements were made in several laboratories, using a majority of the existing techniques. The solar absorptance and thermal emittance of a surface may be measured directly or indirectly. The reflectance of the surface may be determined and from Kirchhoff's relationship one can calculate the required parameter. Figure 4.4.1 shows the spectral distribution of the extraterrestrial sun and the spectral distribution of the blackbody at different temperatures. About 97% of the solar flux is contained in spectral region ranging from 0·3 to 3.0 μm but the corresponding range for the blackbody radiation distribution at 300°K is from 4 to 60 μm. The blackbody radiation distribution contains a significant amount of radiation flux in wavelength region beyond 20 μm. Therefore, for the perfect measurement of emittance, reflectance measurements up to 60 μm are necessary.

4.4.1 *Indirect Determination of Solar Absorptance and Thermal Emittance from Reflectance Data*

The spectral reflectance can be easily measured in the UV, visible, and near infrared region by using commercially available spectrophotometers, but two primary difficulties are associated with these spectrophotometers. One is that most of the devices are comparison devices, i.e., they determine the reflectance of the sample by comparing it with a standard reflector. The second is that the reflectance can be determined only at some particular incidence angle, usually nearly normal. The absolute measurement of reflectance of standard reflector is necessary. Magnesium oxide is frequently used as standard reflector, but its spectral reflectance is subject to variation (159).

Several authors (160–168) follow a number of approaches to avoid the requirement of standard reflector. Gates et al. (160) use a varying number of multiple reflections from identical specular surfaces and determine the absolute spectral reflectance over a range of angles (20–60°) with an accuracy of 2%. Bennett and Kachler (161) describe a precision method (0.1%) for measuring the square of the absolute reflectance of specular surfaces near normal incidence with a Strong type reflectometer (162). The detailed principle of the reflectometer is discussed by Drummeter and Hass (163). Edwards et al. (164) describe and analyze an integrating sphere reflectometer which is not dependent on a standard reflector and also permits the variation of angle of incidence. Harris and Fowler (165), using a technique similar to that of Gates, have determined the reflectance of gold from 8.5 to 84 μm.

Directional reflectance can be measured in a direct manner by irradiating a sample from a particular direction and then detecting all reflected radiation regardless of its directional distribution. Another approach is to radiate the specimen perfectly diffusely and detect the reflected intensity in a small solid angle. This approach is referred as "reciprocal determination."

Direct total reflectance measurements have been made with Coblentz sphere (166, 167), but this method gives some erroneous values due to a characteristic decrease of reflectivity of metals in the short wavelength region of solar spectrum. In addition, it yields erroneous values for diffuse materials due to nonuniform angular response of the detector, optical aberrations, and interreflections (168).

Dunkle et al. (169) describe an instrument which permits rapid direct measurement of reflectance of materials irradiated with solar energy that is transmitted through the atmosphere. The apparatus consists of an equatorially mounted integrating sphere with a thermopile detector. A coating of magnesium oxide 0.03 to 0.04 in. thick was smoked onto polished copper which had been painted with three coats of titanium oxide pigment in a lacquer vehicle. Figure 4.4.2 shows the spectral reflectance of metals with various surface coatings measured by Dunkle et al. (169). The integrating sphere reflectometer gives unsatisfactory performance beyond 2 μm because of the lack of an adequate diffuse reflectance coating for longer wavelengths.

McNicholas (170), Kerrer (171), Tingwaldt (172), Toporets (173), and Edwards et al. (164) describe integrated spheres for direct measurement of spectral directional reflectance in the 0.3–2.5 μm range. Figures 4.4.3 and 4.4.4 show diagrams of these integrating spheres. The shape factor from any receiver element of a sphere to any source element on a sphere is simply the source element area divided by the total area of the sphere. It follows that if the sphere wall is perfectly diffuse and if the detector views all parts of the sphere equally well, the sphere can act as a 2π steradian detector for a specimen anywhere within the sphere.

Gier et al. (165), Tingwaldt (174), and Dunkle et al. (175) describe heated cavity reflectometer for directional reflectance in the 2–20 μm range. The

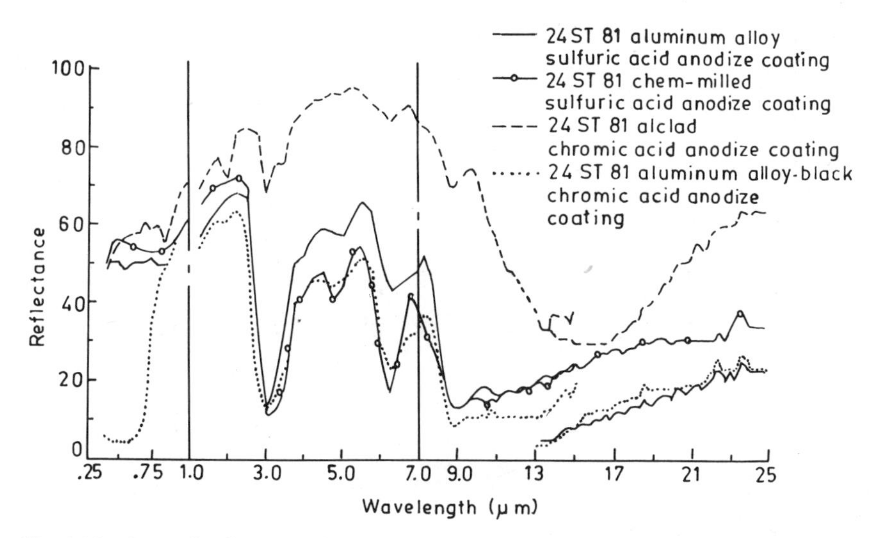

Fig. 4.4.2 Spectral reflectance of various metal surfaces measured with integrating sphere. Adapted from Dunkle et al. (1960) (Ref. 169).

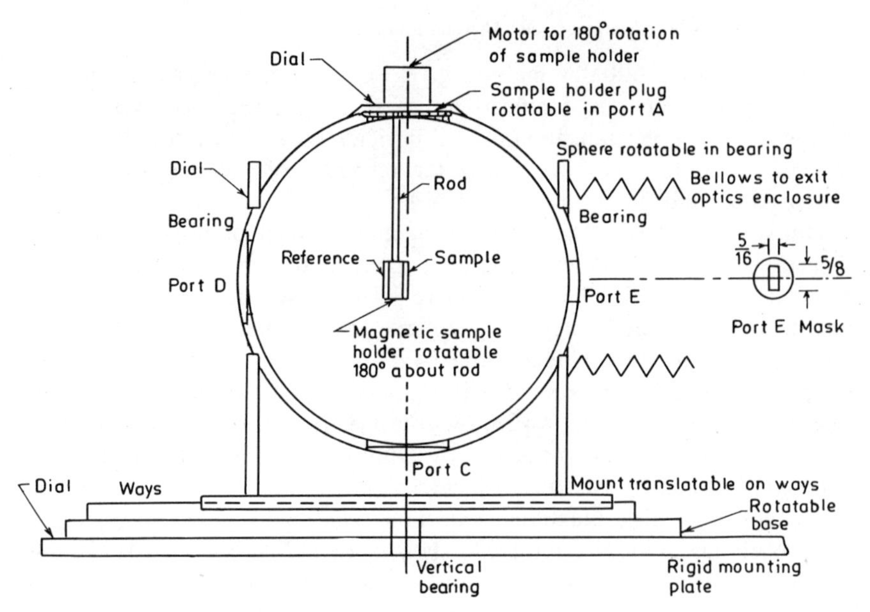

Fig. 4.4.3 Details of directional integrating sphere described by Edwards et al. Adapted from Edwards et al. (1961) (Ref. 164).

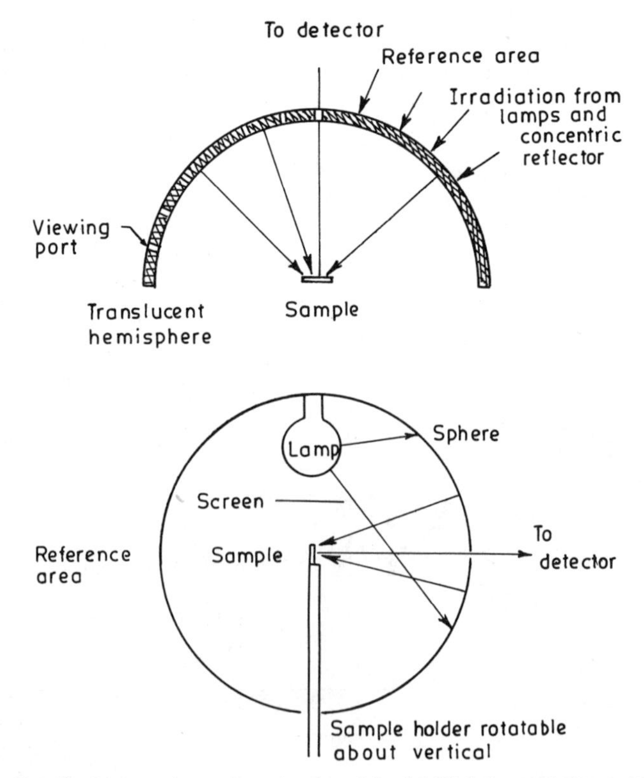

Fig. 4.4.4 Detail of integrating sphere developed by McNicholas and Tingwaldt. Adapted from Tingwaldt (1955) (Ref. 174).

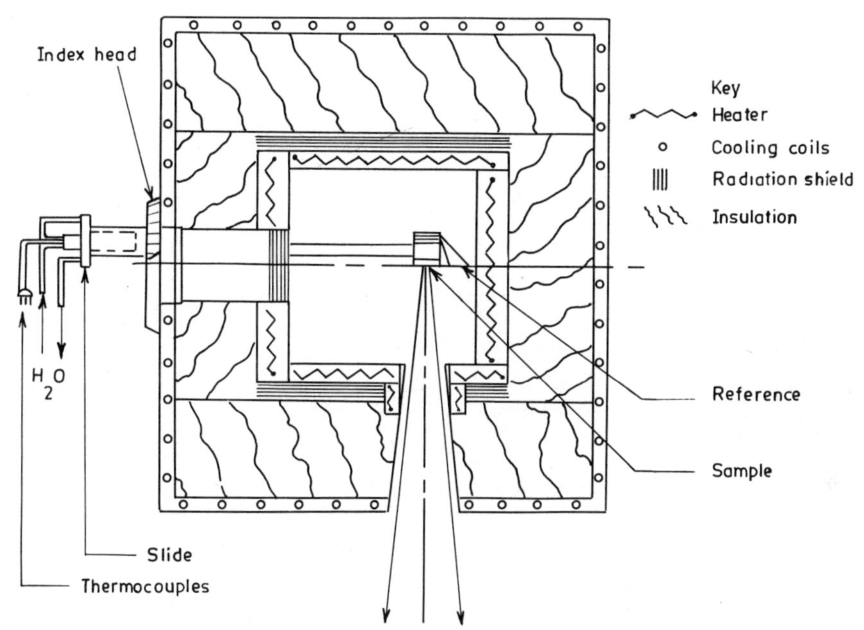

Fig. 4.4.5 Schematic diagram of heat cavity reflectometer. Adapted from Dunkle et al. (1962) (Ref. 175).

mounting of the sample at the end of the rotatable rod within the cavity or sphere permits the polar angle to be varied (174, 175). Figures 4.4.5 and 4.4.6 show diagrams of single-beam heated cavity reflectometer with a rotable sample (175) and Fig. 4.4.7 shows the reflectance spectra of silver film measured with the instrument. The heat cavity gives good results in the intermediate wavelength range from 2 to 20 μm. Gier et al. (176) use two

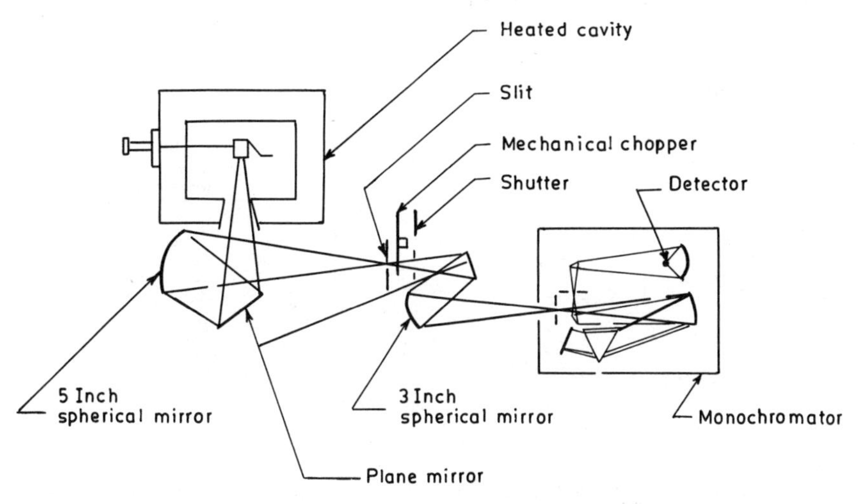

Fig. 4.4.6 Details of heat cavity reflectometer. Adapted from Dunkle et al. (1962) (Ref. 175).

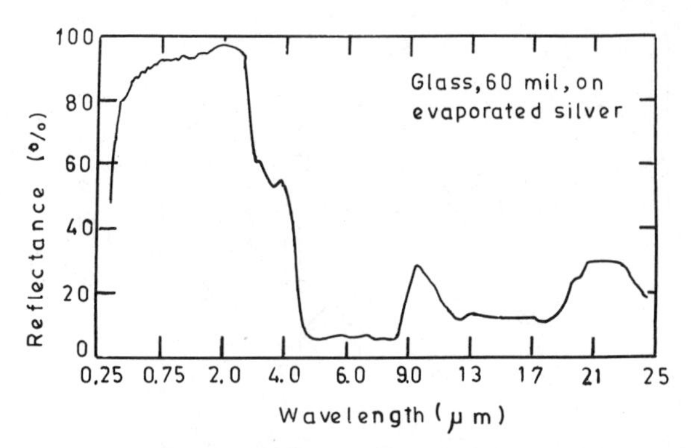

Fig. 4.4.7 Spectral reflectance of silver film on glass measured with integrating sphere and heat cavity reflectometer. Adapted from Dunkle et al. (1962) (Ref. 175).

identical 2π steradian paraboloids and Dunn et al. (177), (178) use an ellipsoid in the direct mode. In the direct mode, radiation is brought onto the sample directionally through a hole in the mirror or via a reflection from a spot on the mirror collected by the 2π steradian mirror, focused on a detector. Neher and Edwards (179) used 2π steradian paraboloid reflectometer in reciprocal mode for the measurement of directional reflectance in the far infrared region. Figure 4.4.8 shows a schematic diagram of the apparatus fabricated by Neher and Edwards (179) and Edwards (180).

Recently Seraphin (181) developed a high-temperature reflectometer for measuring the reflectance of selective surfaces at high temperatures. Figure 4.4.9 shows the photograph of the apparatus. The time required for single spectral run at elevated temperature is much more than that of room-temperature measurements.

Willey (182) developed a portable instrument to determine the solar

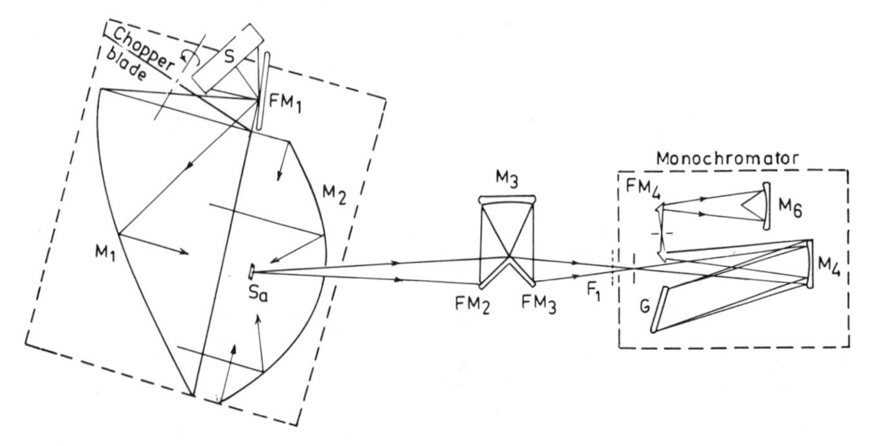

Fig. 4.4.8 Schematic diagram of reciprocal paraboloid reflectometer. Adapted from Neher and Edwards (1965) (Ref. 179).

Fig. 4.4.9 Photograph of high-temperature reflectometer developed by Seraphin. Adapted from Seraphin (1974) (Ref. 181).

absorptance of flat, opaque materials. The reflectometer consists of an approximately elliptical radiation collector coated with white reflecting paint together with imaging lens, a sample measurement port and a silicon detector, as shown in Fig. 4.4.10. The active area of the detector views the wall of the reflector and is positioned so that radiation reflected directly from the sample is not incident on the active area. The light source for the reflectometer is a projector lamp. The measurement procedure consists of first determining a zero reading (V_0) with the measurement port uncovered, a 100 percent reading (V_{100}) with the standard over the measurement port and finally the reading from the sample (V_s). The fractional reflectance of the sample R_f is

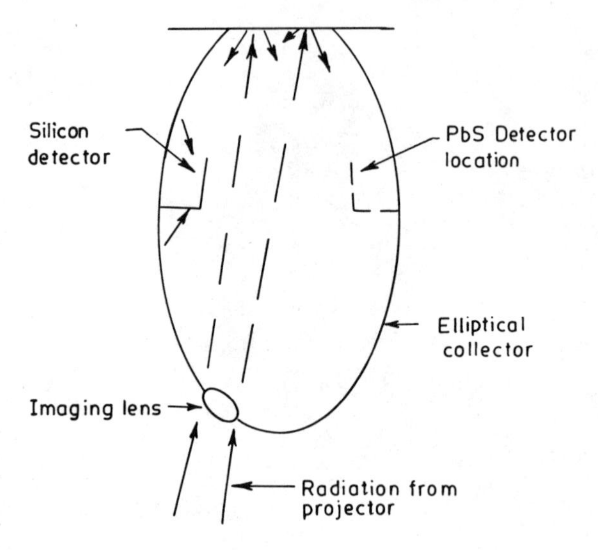

Fig. 4.4.10 Diagram of the Wiley Alpha meter reflectometer. Adapted from Pettit (1978) (Ref. 199a).

calculated as

$$R_f = \frac{V_s - V_0}{V_{100} - V_0} \tag{4.4.1}$$

The true reflectance may be calculated from

$$\rho = \frac{2.54\, R_f}{1.54 + R_f} \tag{4.4.2}$$

The reflectometer uses a silicon photocell for the detector. Two different type of projector lamps—a Sylvania type DFW tungsten coil filament lamp and a GE quartz line ELH tungsten halogen lamp mounted in an

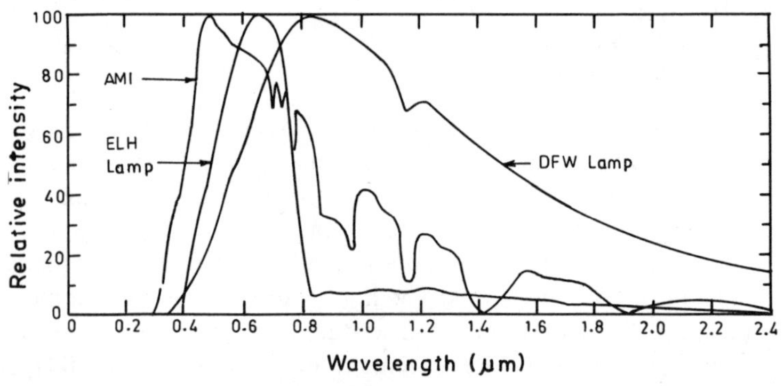

Fig. 4.4.11 Spectral irradiance of sun at AM1 and normalized spectral irradiance curves for DFW lamp and ELH lamps. Adapted from Pettit (1978) (Ref. 199a).

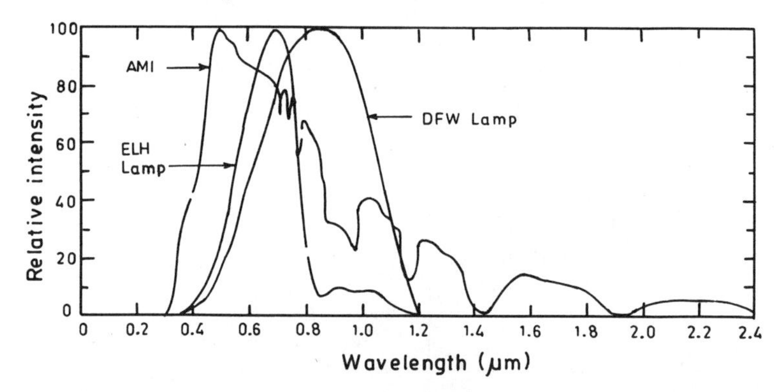

Fig. 4.4.12 Spectral irradiance of DFW lamp and ELH lamps measured with silicon detector and comparison with AM1 sun spectrum. Adapted from Pettit (1978) (Ref. 199a).

ellipsoidal reflector—can be used. Figure 4.4.11 shows a comparison between the AM1 solar spectrum and normalized spectral irradiance curves. DFW lamp spectral distribution has more infrared radiations relative to an AM1 solar spectrum and peaks at 0.8 μm. Both the projector lamps have less amount of the UV part relative to the AM1 spectrum. The combined response of the detector with the measured spectral irradiance gives a spectrum as shown in Fig. 4.4.12. The silicon detector response curve peaks at 0.9 μm and has a sharp cutoff at 1.1 μm. Therefore the reflector is insensitive to the 25% of the AM1 solar spectrum measured by using silicon detector. For improvement, a PbS detector was used and placed inside the ellipsoidal radiation collector of the reflectometer at a point symmetrically opposite to the silicon detector. The detector response curve of PbS peaks at 2.1 μm. The reflectance of $MgCO_3$ standard at 2.1 μm is approximately 0.75 and so the fraction reflectance values, R_f, were divided by 0.75. The overall accuracy of the reflectance values measured with the PbS detector is probably $\pm 15\%$ because of linearity, stability, and standard limitations in the present arrangement. The solar average reflectance values for black chrome samples were calculated by weighing the reflectance measured with the silicon detector by 75% and the reflectance measured with the PbS detector by 25%. The final solar absorptance values are shown in Fig. 4.4.13 as a function of true values. The average difference in α_s as determined by Willey instrument and true α_s values is only ± 0.012 absorptance unit, with the maximum error being ± 0.03 absorptance unit.

4.4.2 *Direct Determination of Thermal Emittance and Solar Absorptance*

Absorptance and emittance are usually calculated from reflectance measurements made with reflectometers, as described in previous section. To determine emittance at a particular wavelength and in a particular direction,

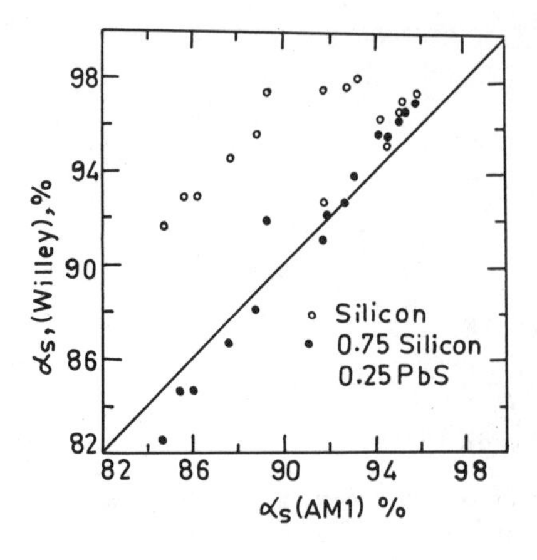

Fig. 4.4.13 Comparison between the solar absorptance with the solar absorptance calculated from absorptance values measured with both Si and PbS detectors. Adapted from Pettit (1978) (Ref. 199a).

usually nearly normal, integrated spheres may be used (169). For the exact determination of hemispherical emittance of anisotropic surfaces (having an angular reflectance distribution), a simpler and more direct approach is required. Several authors have used various techniques (175, 179, 183–187); some of these were used for absolute measurement of emittance and others were used to obtain measurements quickly at the sacrifice of accuracy. The calorimetric or radiometric technique is considered to be more accurate and is widely used for the determination of total hemispherical emittance and α/ε ratio. For a calorimetric measurement, a sample of known radiating surface area is placed in a controlled environment where heat loss or gain is by radiant heat transfer only. The emittance is then computed from the amount of power required to maintain the sample at measured equilibrium temperature. This is known as steady state technique. With a knowledge of the heat capacity of specimen, the emittance may be computed from the rate of temperature change of the sample. This is known as "transient technique." If the input power is supplied directly from sun or solar simulator, then both solar absorptance and emittance as a function of temperature can be obtained as discussed by Fussell et al. (188).

The radiant heat exchange between two bodies, where body 1 is completely surrounded by body 2, is characterized by the heat flow out of body 1 and given by (189):

$$Q_{\text{out}} = \frac{\sigma \varepsilon_1 A_1 (T_1^4 - T_2^4)}{1 + \varepsilon_1 A_1 (\varepsilon_2 A_2)^{-1} (1 - \varepsilon_2)} \tag{4.4.3}$$

where σ is Stefan–Boltzmann consant, A_1, A_2, T_1, and T_2 are surface areas and temperatures, and ε_1 and ε_2 are hemispherical emittances of bodies 1 and 2, respectively. If the surface area ratio of body 2 and body 1, i.e., A_2/A_1 is very large, i.e., $A_2/A_1 \to \infty$ and emittance of bodies tends to unity, eq. (4.4.3) reduces to

$$Q_{\text{out}} = \sigma \varepsilon_1 A_1 (T_1^4 - T_2^4) \tag{4.4.4}$$

If a power P_1 is being supplied to the sample or body 1, then in equilibrium $P_1 = P_{out}$, where P_{out} is the power output. Therefore by measuring the equilibrium temperature surface area of the sample and the temperature of the environment, one may compute the emittance of body 1 using eq. 4.4.4. If the body 1 is not in equilibrium but is uniform in temperature, then the rate of change of temperature of the sample is expressed as (188):

$$mc_p \frac{dT}{dt} = P_1 - P_{out} \tag{4.4.5}$$

where m is sample mass, c_p is specific heat of the body 1 and t is time. Thus by knowing heat capacity of body 1 one may compute the emittance from measuring the rate of temperature change in cooling process. The major disadvantage in former technique (steady state technique) is the large time constant and correspondingly long times required to reach equilibrium.

By inserting eq. 4.4.4 in eq. 4.4.5 one gets:

$$mc_p \frac{dT}{dt} = P_1 - \sigma \varepsilon_1 A_1 (T_1^4 - T_2^4) \tag{4.4.6}$$

At equilibrium conditions $dT/dt = 0$, i.e., the equilibrium temperature can be written as

$$T_{eq} = \left(\frac{P_1}{\varepsilon_1 \sigma A_1} - T_2^4 \right)^{1/4} \tag{4.4.7}$$

Equation 4.4.6 can be rewritten as

$$mc_p \frac{dT(t)}{dt} = \sigma \varepsilon_1 A_1 (T_{eq}^4 - T_{(t)}^4) \tag{4.4.8}$$

The solution of above equation for $T(t)$ is given by (189)

$$T(t) = T_{eq} - Ce^{-t/\tau} \tag{4.4.9}$$

where C is a constant and

$$\tau = mc_p / 4\varepsilon_1 \sigma A_1 T_{eq}^3 \tag{4.4.10}$$

Thus the sample approaches the equilibrium exponentially with time and the time constant depends upon the sample parameters as expressed in eq. 4.4.10.

Pettit (189) developed a technique called the "delta power calorimetric emittance technique" by which several equilibrium positions can be achieved in a short period of time in low-emittance materials. Figures 4.4.14 and 4.4.15 show schematic diagrams of the apparatus used for calorimetric emittance measurements developed by Pettit. The sample was placed in vacuum chamber (to eliminate gas convection and conduction losses) as well as inside in isothermal chamber which is at a temperature much lower than the sample temperature. The liquid-nitrogen-cooled copper chamber was

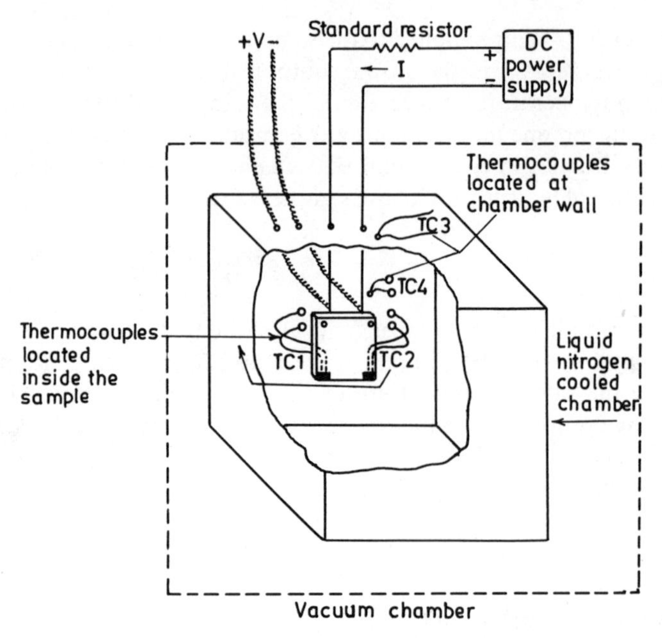

Fig. 4.4.14 Schematic diagram of apparatus used for calorimetric emittance measurements developed by Pettit. Adapted from Pettit (1978) (Ref. 189).

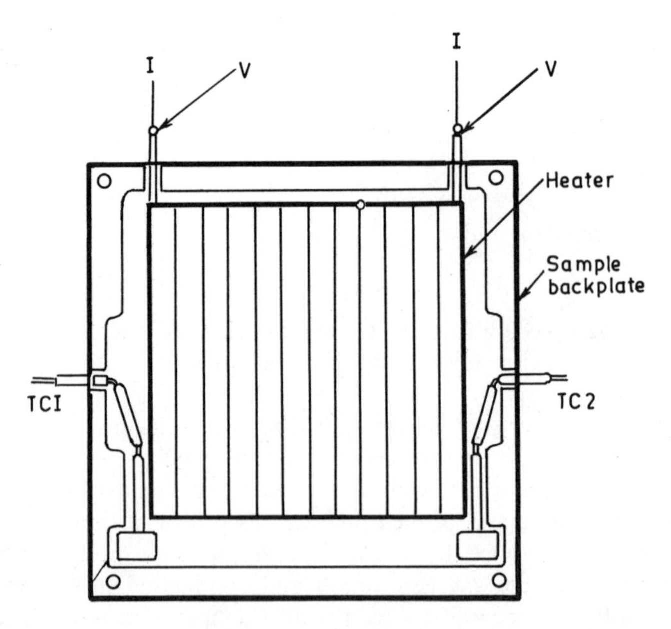

Fig. 4.4.15 Cross section of the emissometer. Adapted from Pettit (1978) (Ref. 189).

painted on the inside with black velvet paint [$\varepsilon_{t,H} \geq 0.88$ from 77°K to 300 °C (190)]. The sample was composed of two parts (1) a back plate which has a machined cavity 0.085 in. deep to accommodate an electric heater together with two chromel-alumel thermocouples and (2) a flat cover plate fitted with four countersunk holes to accommodate flat-head machine screws. Both the heater and inside surface of each sample were painted with pyromark paint [$\varepsilon_{t,H} \geq 0.80$ from 250°C to 3000°C (191)] to minimize thermal radiation heat transfer between the heater and the sample.

Sadler et al. (192) developed a device for total hemispherical emittance which has been used at Grumman Aircraft Engineering Corporation. The emissometer is based on the principle of guarded heater and is shown in Figs. 4.4.16 and 4.4.17. The guard provides isothermal surroundings for the test specimen and its heater so that heat is lost only by radiation from the sample surface which faces the cold wall in the vacuum chamber. The guard heater cup was constructed of heavy aluminium to prevent temperature gradients along its inside surface and fibreglass legs support the specimen heater so that the conduction errors are minimized. The test specimen was attached to the heater with a thin layer of silicone grease to ensure good thermal contact. The heating elements of the specimen and guard heaters were constructed of stainless-steel jacketed nichrome wire. All thermopiles and thermocouples were made of calibrated copper and constantan wire to

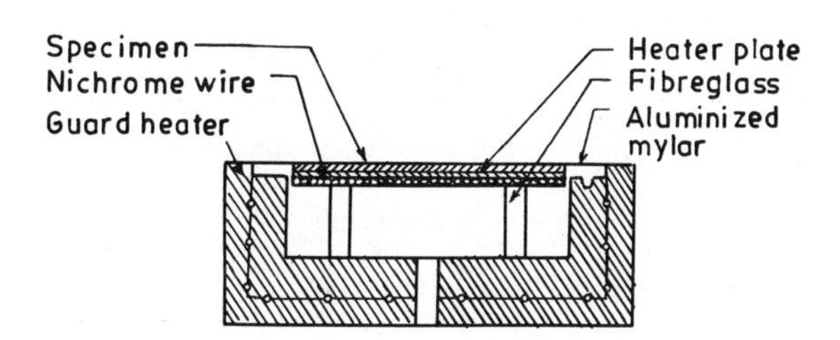

Fig. 4.4.16 Cross section of the emissometer developed by Sadler et al. Adapted from Sadler et al. (1963) (Ref. 192).

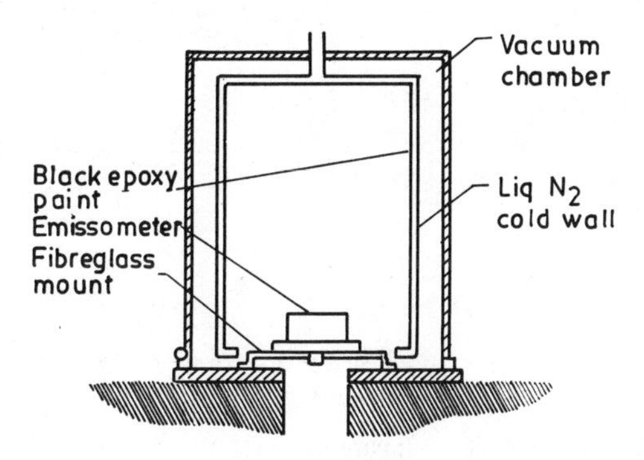

Fig. 4.4.17 Emissometer in bell jar. Adapted from Sadler et al. (1963) (Ref. 192).

ensure adequate sensitivity in the range of the test temperatures. In the steady state condition, all of the power supplied to the specimen heater must be lost as radiation from the specimen surface to the cold wall. Therefore from the exact measurement of power supplied to the specimen heater, the quantity of heat emitted from the test surface to its surroundings can be calculated. A comparison of this quantity with the amount of heat emitted by a black surface of the same area determine the hemispherical emittance of the specimen. The hemispherical emittance may be calculated as (192)

$$\varepsilon_{t,H} = \frac{C\, I_s^2\, R_s}{\sigma(T_1^4 - T_w^4)A_s} \tag{4.4.11}$$

where R_s is the specimen heater resistance, I_s is the current supplied to heater, C is a constant T_1 is the temperature of the specimen, and T_w-temperature of cold wall.

Zerlaut (193) developed an apparatus for the measurement of the total normal emittance of surfaces at elevated temperatures. The total normal emittance was measured as the ratio of the thermoelectric EMF generated by the sample to the thermoelectric EMF generated by the blackbody at the same temperature and pressure. For a detector, a well-baffled, 28-junction, iron–constantan, radial thermopile was built. Figure 4.4.18 shows schematic diagram of the emissometer developed by Zerlaut (193). Gaumer and Stewart (194) and Willrath and Gammon (195) developed a solar calorimeter for the measurement of thermal emittance anb the α/ε ratio. The α/ε device operates

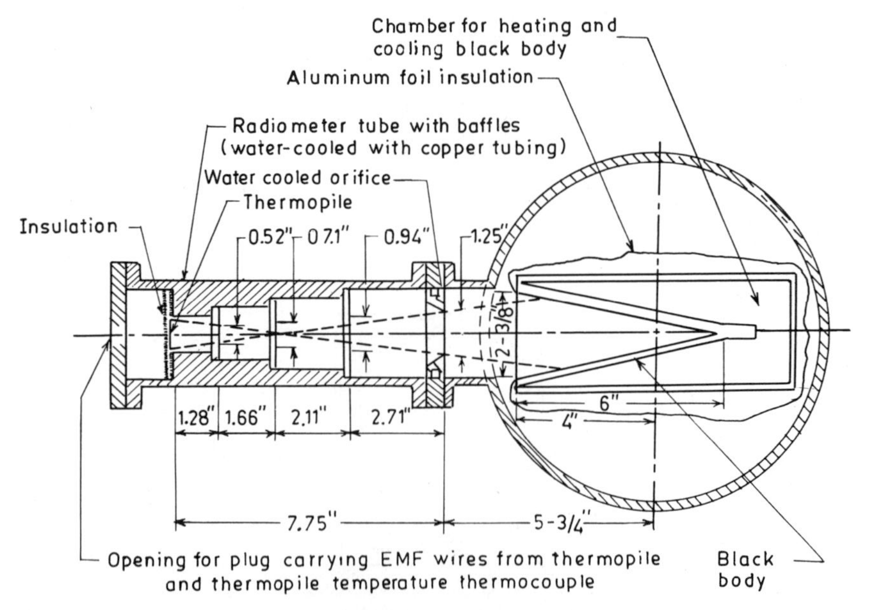

Fig. 4.4.18 Schematic diagram of the apparatus used for the normal emittance measurements at elevated temperature. Adapted from Zerlaut (1963) (Ref. 193).

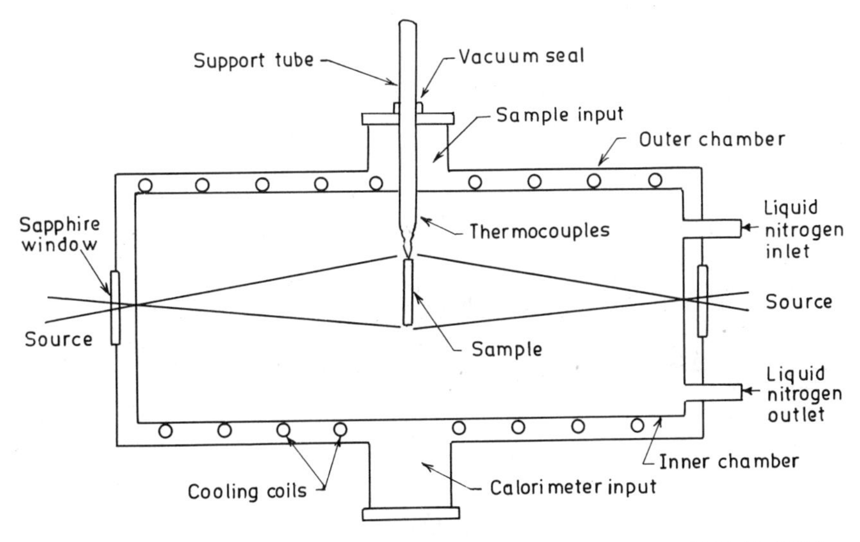

Fig. 4.4.19 Schematic picture of α/ε device used by Gaumer and Stewart (1963) (Ref. 194).

on the principle that a body thermally isolated in vacuum will absorb incident radiant energy according to its absorptance for that radiation, and it will emit thermal energy according to its total hemispherical emittance ε. Figure 4.4.19 shows the schematic diagram of α/ε device (194). Under equilibrium conditions, a balance is established for a specimen absorbing and emitting radiant energy in vacuum, the α/ε ratio may be represented by (194) eq. 4.4.12:

$$\frac{\alpha}{\varepsilon} = \left(\sigma \frac{A_2}{A_1}\right) \frac{T^4}{I} \tag{4.4.12}$$

where A is emitting area, T is sample temperature, and I is the intensity of incident radiation. The intensity of incoming radiations can be determined by placing the calorimeter at sample place and may be expressed as

$$I = K\, T_c^4 \tag{4.4.13}$$

where K is appropriate constant for the calorimeter and T_c is the temperature of the calorimeter. Therefore if the emitting and absorbing areas, i.e., A_1 and A_2 are equal, then

$$\frac{\alpha}{\varepsilon} = \left(\frac{T}{T_c}\right)^4 \tag{4.4.14}$$

Also, under equilibrium conditions when the source is turned off the emittance may be calculated as (194)

$$\varepsilon = \frac{mc_p}{3\sigma A(t_2 - t_1)} \left(\frac{1}{T_1^3} - \frac{1}{T_2^3}\right) \tag{4.4.15}$$

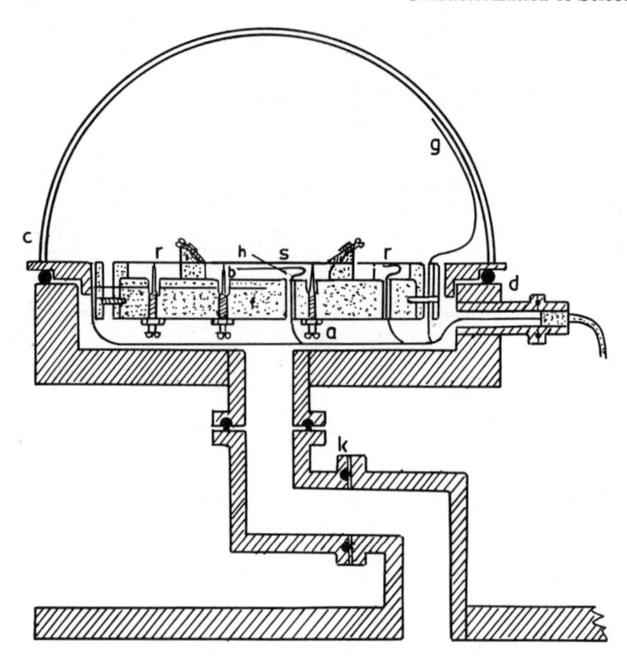

Fig. 4.4.20 Cross section of solar calorimeter. (a) sharp point (b) polished aluminium surface (c) hemispherical glass dome (d) O-ring seal (e) glass points (h, g, f, i) thin copper and constantan wire (j) cylindrical brass base (k) rotating joint. Adapted from Willrath and Gammon (1978) (Ref. 195).

where T_1 is the sample temperature at time t_1, T_2 is the sample temperature at time t_2, and A is emitting area of the specimen.

Figure 4.4.20 shows a schematic diagram of solar calorimeter recently developed by Willrath and Gamman (195) for solar absorptance and thermal emittance measurements. The detailed calculations for solar absorptance and thermal emittance are described by these authors. Figure 4.4.21 shows the absorptance and emittance as a function of temperature for polished copper and black chromeplated copper. McKenney and Beauchamp (196) also developed a solar calorimeter for the direct measurement of solar absorptance and thermal emittance. This apparatus is quite similar to that of Willrath and Gammaon (195). The environment inside the chamber is maintained at ambient temperature and a heat-flow sensor is placed between the sample and heated black.

Recently McDonald at NASA–Lewis Research Center and Willey Corporation (197) have developed portable instruments for solar absorptance and thermal emittance measurements. The instruments can be easily transported and set up at locations where the selective coating is being applied. In addition to this, measurements can be performed in the field to check for coating degradation and/or uniformity both before and after installation.

The Willey ambient emissometer (Model 2158) is a portable instrument

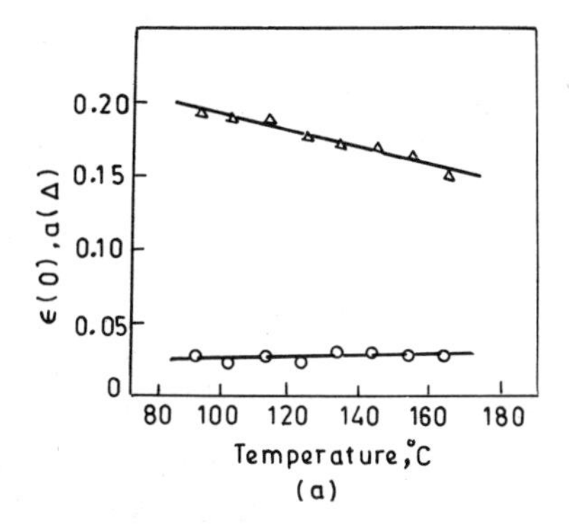

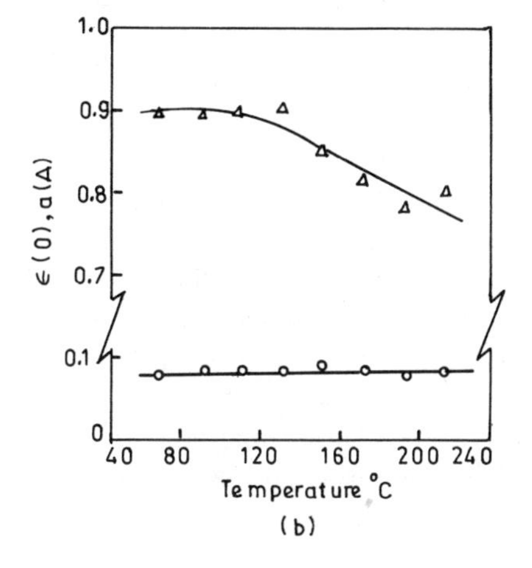

Fig. 4.4.21 Solar absorptance (Δ) and thermal emittance (○) versus temperature of (a) polished copper and (b) black chrome on Copper. Adapted from Williath and Gammon (1978) (Ref. 195).

which consists of a measurement head which is heated to 100°C, and separate display unit. The head consists of an aluminum ellipsoidal reflector cavity with a thermocouple detector centered at the bottom of the ellipsoid close to one focus and a measurement port located near the other focus. The display unit reading is proportional to the temperature difference between the thermocouple detector and the reference thermocouple. The operation principle depends upon the radiation exchange between the thermocouple detector and a sample placed over the measurement port. Thermal radiation from the inside of the ellipsoidal reflector and the sample is incident on the thermocouple detector. If the sample has a high infrared reflectance (low emittance), the temperature of the detector will be very close to the temperature of the measuring head; however if the sample has low infrared reflectance (high emittance) the temperature of the detector will be below the temperature of the measuring head. Figure 4.4.22 shows a schematic diagram of the Willey ambient emissometer. McDonald (197) used parabolic reflector (5 cm in diameter) instead of ellipsoidal reflector, and the thin-foil detector is placed at the focus of parabola and parallel to the axis of the

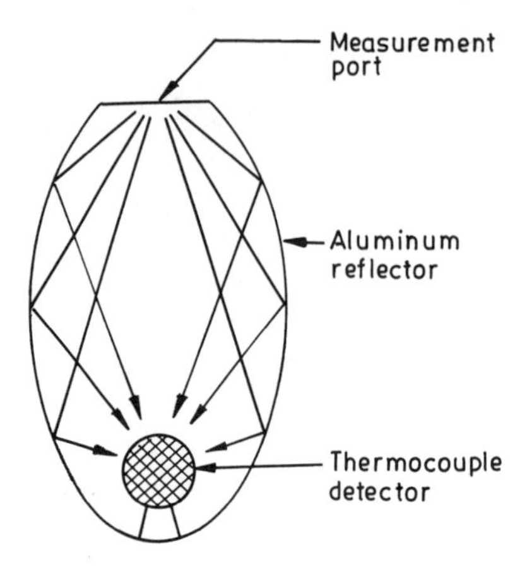

Fig. 4.4.22 Schematic diagram of Willey Ambient emissometer. Adapted from Willey (1976) (Ref. 199).

parabola. The detector foil is heated by radiation, conduction, and convection. Conductive heating is made small by supporting the detector foil only by thin thermocouple wires. Convective heating of the foil could be reduced by placing a thin plastic film across the open face of the reflector. However, this would reduce the signal intensity due to absorption of IR radiation by the plastic film, thereby reducing the sensitivity of the instrument. A schematic diagram of McDonald emissometer is shown in Fig. 4.4.23. For calibration, the high-emittance standard is 3M Nextel Black paint ($\varepsilon = 0.95$) and the low-emittance standard is aluminium foil ($\varepsilon = 0.060$). These emissometers described above, are comparative devices which permit calculations of emittance from comparative measurements of radiant energy for a sample of known emittance and another sample of unknown emittance. Therefore, the accuracy depends upon the accuracy with which the emittance of the standards is known. Recently Willey (198, 199) introduced the Willey 318S Fourier Transform Spectrophotometer which is commercially available. This device

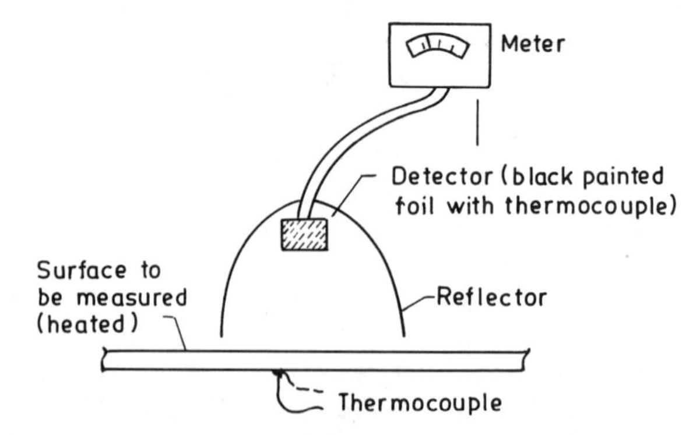

Fig. 4.4.23 Schematic diagram of McDonald's emissometer. Adapted from McDonald (1977) (Ref. 197).

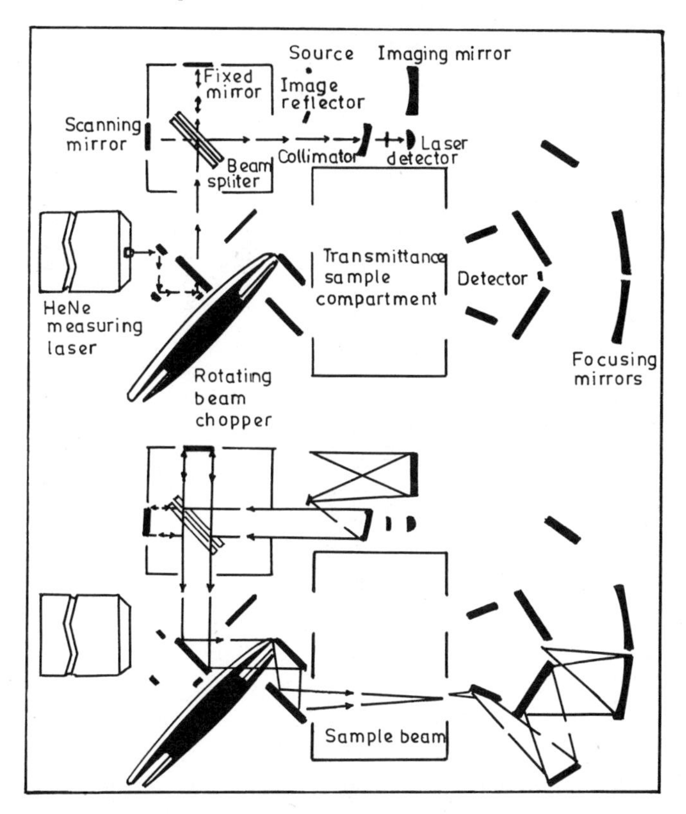

Fig 4.4.24 Optical scheme of Willey 318S Fourier transform spectrophotometer. Adapted from Willey (1976) (Ref. 199).

can rapidly measure the normal reflectance and hemispherical emittance. Figure 4.4.24 shows optical scheme of the instrument. The source is a Nernst glower with platinum heater coils on an automatic starting circuit. The spectral modulator is a Michelson interferometer which essentially causes different wavelengths to be modulated at different time frequencies. Normal diffuse reflectance is measured with the sample placed over the sample port and a reference placed over the reference port. The specular gold mirrors, specular aluminium mirrors, and diffuse gold surfaces are used as references. The normal reflectance, and thereby normal emittance ($\varepsilon = 1 - r$), can be measured. The spectral emittance is measured by the blackbody spectrum at a particular temperature and integrated to give the relative directional emissivity. A special procedure was developed for the Willey 318S to measure hemispherical emittance by hemispherically illuminating the sample and measuring the light that is reflected hemispherically. The usual sample and reference ports are covered with diffuse reflectors similar to the wall of integrating sphere. Three spectra are taken: (1) with the sample covering the spectral port, (2) with the spectral port open, and (3) with the

Table 4.4.1 Comparison of Techniques for Measuring Solar Absorptance and Thermal Emittance

Serial No.	Techniques	Wavelength Range (μm)	Angles (degrees)	Major Features	Typical Errors
1.	Integrating sphere	0.3–2.5	nearly normal	The sphere acts as a 2π steradian detector for a specimen in the sphere, performance is excellent up to 2.0 μm due to diffuse reflectance coating of $MgCO_3$	Direct radiation error, directional sensitivity error, stray radiation interreflections, nonlinearity of detection system, nonuniformity of sphere wall coating, inter-reflections with sample
2.	Heated cavity	2–20	7–75	2π off-axis illumination, uncertainties in cavity temperature uniformity, cavity opening, and reference temperature.	Sample emission, temperature gradients, interreflections, stray radiations, nonlinearity
3.	2π-steradian mirror	5–80		2π steradian mirror has source side chopping, in the direct as well as the reciprocal mode of operation, measurements with specimen temperatures at either elevated or cryogenic levels	Optical aberrations, directional sensitivity of detector or source, variation of mirror reflectance with angle and position, radiation, and nonlinearity

4.	Paraboloid	2–75	17	2π steradian illumination chopped uniform source, gold reference difficult for angular use, physically large components	Presence of exit port having zero radiance, variation in phase of chopping beams to mirror M_1, interreflections between the sample and source, fluctuating sample emission
5.	Ellipsoid	2.5–80	5–79	2π steradian, absolute method, vacuum or inert gas operation, hemispherical source chopper, point by point operation, initial high cost	Angular and spatial sensitivity of the detector, aberration in optics, losses due to the entrance hole in the ellipsoidal mirror, etc.
6.	Specular	2.5–50	10	Relative method, uses commercial attachment with front surfaced aluminized mirror reference	Surface irregularities, nonuniform reflections
7.	Portable Willey Alpha meter	0.3–2.5	nearly normal	Measures normal solar reflectance, compact, relative measurement, low cost	Spectral irradiance spectrum of light source, directional characteristics of detectors, etc.

Table 4.4.1 (*Continued*)

Serial No.	Techniques	Wavelength range in μm	Angles (degrees)	Major Features	Typical Errors
8.	Calorimetric steady state technique.	All	All	Absolute direct measurement of hemispherical emittance, Resistance heating of sample, heat balance in liquid nitrogen isothermal and evacuated chamber, uncertainties in temperature measurement, long time required for single operation	Thermal conduction loss and/or input from the thermocouple, voltage and heater wires, gas conduction and convection losses, thermal gradient in the sample.
9.	Calorimetric Transient Technique	All	All	Temperature decay in vacuum, uncertainties in heat leak, less time is required for operation than for steady state technique	Same as above
10.	Solar calorimeter	All	All	Direct measurement of thermal emittance and α/ε ratio, optical measurements may be made at ambient and elevated temperatures, sun radiation solar simulator may be used, instead of AM1	Heat loss from the top of the sample, heat loss by conduction through the thermocouples, spectral distribution of temperature in space over the dome, etc.

specular port covering a portion similar to the sphere wall. With these three spectra and the dimensions of the sphere and ports, the reflectance of the sphere wall and hemispherical reflectance of the sample can be computed. With the Willey 318S spectrophotometer, it is practical to measure the reflectance or emittance of most materials at ambient temperature or even at elevated temperature. Table 4.4.1 gives a comparison of techniques for measuring solar absorptance and thermal emittance.

5

Black Solar Selective Surfaces

The key requirement of an efficient solar absorbing coating is the spectral selectivity. A surface whose optical properties of absorptance, reflectance, and emittance vary in the solar and thermal IR regions, is termed a spectrally selective surface. For solar energy applications, a spectrally selective surface should capture the maximum solar energy in the high-intensity visible and near IR spectral regions and should have minimum emittance for thermal IR radiations. An ideal spectrally selective surface should have an abrupt transition between the low- and high-reflectance regions around 2 μm, which is approximately the limit of the solar spectrum. In addition to the spectral selectivity, a practical selective surface should be stable at the operating temperature and must have a long life and low fabrication cost. Spectral selectivity can be achieved in a variety of ways. The various types of absorber surfaces are: intrinsic absorbers, absorber–reflector tandems, multilayer interference stacks, powdered semiconductor reflector combinations, optical trapping systems, composite material films, and quantum size effects. Developments in the various techniques are reviewed in this chapter. The key manufacturing processes, the optical characteristics, and thermal stability of each type are analyzed in detail. Since the cost effectiveness is an important consideration in solar energy applications, a knowledge of the various properties helps in the choice of a coating for a particular application. The recent development of semiconductor paints has provided a low-cost large-area process for large-scale applications in solar collector systems and has received particular attention in this chapter. The magnetron sputtering technique for depositing black selective surfaces on solar concentrators for high-temperature applications is a new and significant development and has also been described. The enhancement in the solar collector thermal performance by the application of honeycomb structures between the absorbing surface and transparent cover is also discussed. Finally, a comprehensive table (Table 5.15.1) lists the properties of various selective coatings.

5.1 Solar Selective Absorbing Surfaces

Almost all black selective surfaces are generally applied on the metal base which provides low emittance for thermal IR radiations and simultaneously good heat-transfer characteristics for solar photothermal applications. Solar radiations may be collected as shown in Fig. 5.1.1(a) where most of the energy is absorbed and a small amount of energy reflected and radiated by the surface. Such a surface is called good selective surface. However, if

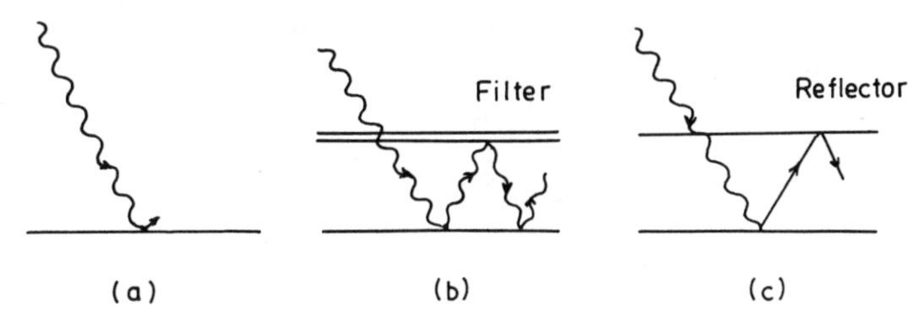

Fig. 5.1.1 Various mechanisms for the absorption of photothermal energy.

Table 5.1.1 Black Selective Coatings

Type	Materials	Reference
Intrinsic Materials	Hafnium carbide	
	Metallic tungsten	
	MoO_3 doped Mo	
	Dieuropium trioxide (Eu_2O_3)	Seraphin and coworkers (83, 200, 201)
	Rhenium trioxide (ReO_3)	
	Divanadium pantaoxide (V_2O_5)	
	Lanthanum hexaboride	
Absorber–Reflector metal tandems	Impure semiconductors	
	Black nickel	Tabor (64, 207–209); McDonald (210)
	Black copper	Hottel and Unger (235); Tabor (207, 209); Mattox and Sowell (242); Mar et al. (211, 220); Driver and McCormik (241)
	Black chrome	Pettit and Sowell (227); Harding (230); Mar et al. (220, 211); McDonald (217–219); Lin et al. (221); Lampert (85)

(contd.)

Table 5.1.1 (*Continued*)

Type	Materials	Reference
	Black iron	Mar et al. (211, 220); Lin et al. (221); Christie (236)
	Cobalt oxide	Vander Leij (251)
	Tungsten oxide	Vander Leij (251, 252); Thronton (254)
	Conversion coatings	
	Copper sulfide	Mattox and Sowell (242)
	Chemically coverted zinc	Encheva (256, 257); McDonald (255); Vander Leij (252)
	Colored stainless steel	Smith (260); Granziera (261); Karlssan and Ribbing (258, 259)
	Alcoa coatings	Powers et al. (266); Cochran and Powers (267)
	Pure semiconductors	
	Silicon, germanium	Seraphin (273, 274); Mattox and Kominiak (271)
	PbS	Williams et al. (278); McMahan and Jasperson (279); Mattox and Kominiak (271); Gupta et al. (282, 283)
	Metal silicide and carbide	Harding (284–288)
Multilayer interference stacks	Dielectric metal dielectric	Seraphin and Meinel (83); Meinel and Meinel (206); Park (297); Mar et al. (304); McKenney et al. (305)
Powered Semiconductor–Reflector combination	Semiconductors	Williams et al. (278)
	Inorganic metal oxide	Mar et al. (220)
	Organic blacks	Lin et al. (221)
	Metal dust pigmented selective paints	Gupta et al. (295); Telkes (294)
Optical trapping systems (dendritic structure)	Tungsten	Tabor (209, 232); Seraphin (83);
	Nickel	Grimmer et al. (318)
	Aluminium	Santala and Sabol (320)
Composite Material	Metal–Insulator films	Cohen et al. (340)
		Granqvist et al. (352); Fan et al. (343)
	Semiconductor–Insulator films	Craighead et al. (358, 359); Gittleman et al. (341)
Quantum Size Effects	InSb on silver and aluminium substrate	Burrafato, Giaquinta, Mancini, and coworkers (369–373)

the particular surface does not have good enough selectivity, it may be enhanced by adding one or more filters [Fig. 5.1.1(b)]. This will allow the incoming solar energy to hit the absorbing surface but will prevent the energy radiated from this surface from escaping. Similarly a reflector may be added which will reflect the energy towards the absorbing surface and thus prevent it from escaping [Fig. 5.1.1(c)]. This arrangement can be repeated any number of times and the surfaces can be stacked one on the top of the other. Selective absorbing coatings can be divided phenomenologically into various classes depending upon the mechanism associated with the functional performance. Table 5.1.1 lists various classes and types of black selective coatings.

5.2 Intrinsic Materials

There is no material occuring in nature which exhibits ideal solar selective properties. There are of course some having approximate selective properties. The intrinsic solar selective properties can be found in two types of materials: (1) transition metals and (2) semiconductors. For each one to serve as an intrinsic absorber would mean that it would have to be greatly modified. In general, metals exhibit a plasma reflection edge at about 0.3 μm, which can be shifted towards the IR by creation of internal scattering centers. Figure 5.2.1 shows spectral reflectance of MoO_3 doped Mo developed by Seraphin and associates (200, 201). For comparison, pure tungsten, which is one of the most wavelength selective metals is also shown in the diagram. On the other hand by making a semiconductor highly degenerate, it may be possible to suppress its plasma frequency in the IR. Seraphin (83, 200, 201) and Ehrenreich (203) discussed several considerations in making intrinsic semiconductors as solar selective surfaces. Hafnium carbide (HfC) has

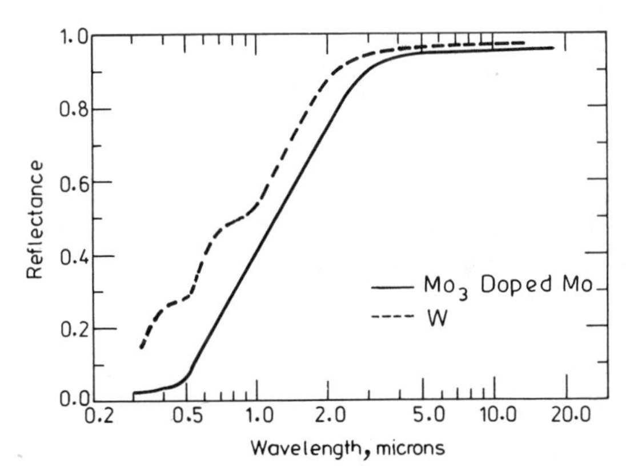

Fig. 5.2.1 Spectral reflectance for MoO_3 doped Mo and metallic tungsten intrinsic materials. Adapted from Seraphin (1979) (Ref. 200).

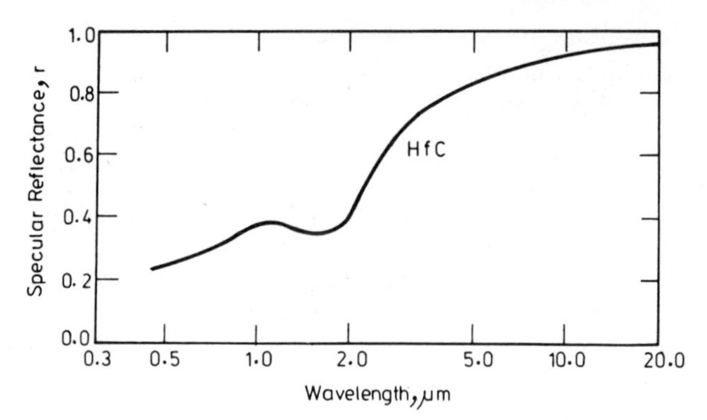

Fig. 5.2.2 Spectral reflectance of hafnium carbide (HfC) intrinsic material. Adapted from Seraphin (1974) (Ref. 83).

different reflectances in different parts of the spectrum. It has high reflectance in the thermal IR region and high absorptance in solar region (83, 202). The reflectance curve for HfC is shown in Fig. 5.2.2. This profile results in a thermal emittance of about 0.10 and solar absorptance of about 0.65. For successful and efficient use in photothermal conversion the solar absorptance should be more than this and the transition from low reflectance to high

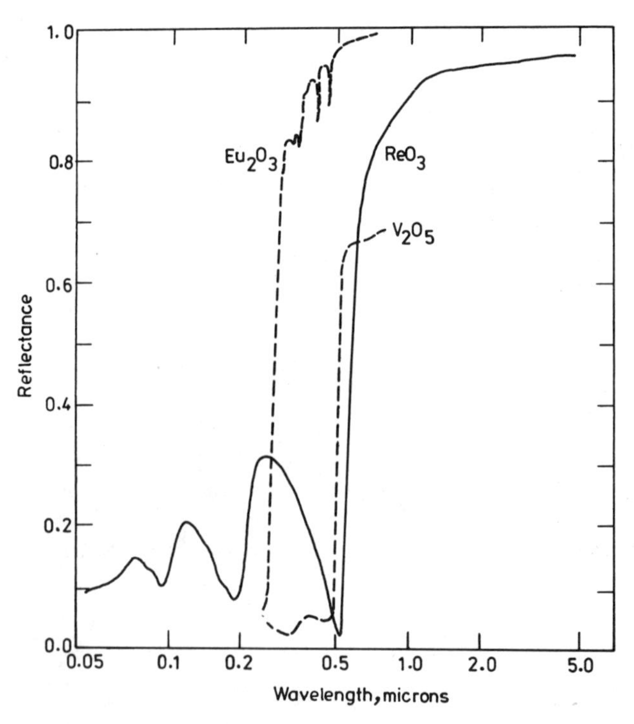

Fig. 5.2.3 Spectral reflectance of dieuropium trioxide, rhenium trioxide single crystal, and divanadium pentaoxide materials. Adapted from Touloukian et al. (1972) (Ref. 204).

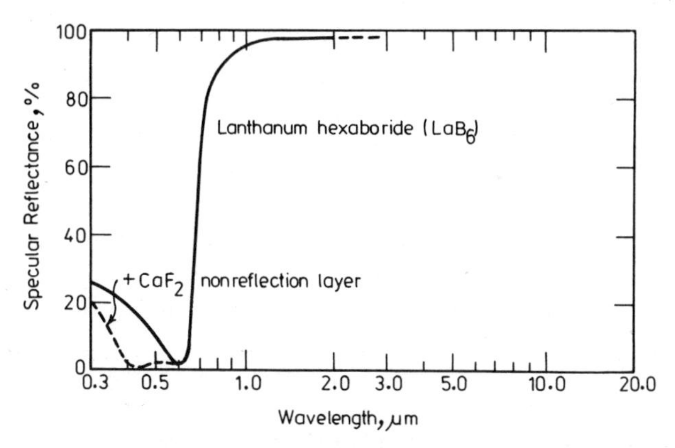

Fig. 5.2.4 Specular reflectances of lanthanum hexaboride on glass and CaF_2 overcoated sample. Adapted from Kauer (1966) (Ref. 205).

reflectance should be steep. Perhaps these characteristics can be achieved either by some structural and/or compositional changes in the lattice of HfC or by coating it with a quarter wave of dielectric material. HfC selective surface is useful as an absorbing surface at elevated temperatures because of its high melting point.

Three transition metal oxides namely dieuropium trioxide, rhenium trioxide single crystal, and divanadium pentaoxide show wavelength selective properties as shown in Fig. 5.2.3. All of these oxides exhibit optical transitions from low reflectance to high reflectance too early in the solar spectrum. The distinct feature of these oxides, as observed by Ehrenreich et al. (203) is an incompletely filled *d* shell of the metallic ions. When this ion is combined with oxygen, the *d* shell electrons become localized, resulting in this optical property. For the further development of intrinsic materials as solar selective surfaces, a thorough understanding of band structure, structural imperfection, etc., is necessary.

Lanthanun hexaboride, LaB_6 (204), is another material which has high IR reflectance in the thermal spectrum and simultaneously high transmittance in the solar spectrum. Therefore LaB_6 acts as selective window or transparent heat mirror because of its high transmission (0.85) in solar spectrum.

The optical transition from high transmission to high reflectance takes place at 0.5–0.6 μm. Figure 5.2.4 shows the spectral reflectance of LaB_6 on glass reported by Kauer (205).

5.3 Absorber–Reflector Tandems

In absorber–reflector tandems, a coating having high absorptance at solar wavelengths (i.e., it is black but is transparent to long wavelength radiation) is deposited onto a highly IR-reflecting metal substrate (e.g., aluminium, copper, silver, etc.). Therefore, the system has high solar absorptance due to the black exterior deposit and low thermal emittance due to the metallic reflector substrate. High absorption of the exterior coating may be either intrinsic in nature or geometrically enhanced, or may be combination of the two. Generally these black coatings are semiconductive in nature and their absorption is a result of interaction of photons having energies greater than the band gap. Therefore, the coatings absorb the photons as a result of raising the materials valence electron into the conduction band and the photons of less than band gap energy are transmitted through the material unaffected.

Seraphin (83, 206) has discussed the basic requirements for absorber–reflector tandems. He assumed an absorber of thickness d and optical constants n and k to be overlaid on a reflector of reflectance R and assumed that the reflection on the two interfaces obeys Fresnel's equation and that no interference takes place between the two reflected beams. By applying Fresnel's equations, Seraphin (83) plotted absorptance as a function of product of absorption coefficient (K) and thickness (d). The profile was calculated for two values of refractive indices ($n = 2$, 4) of the absorber. The absorptance profile is shown in Fig. 5.3.1 as a function of absorptance product

$$kd = \left(\frac{4\pi k}{\lambda}\right)d \tag{5.3.1}$$

It is obvious from the profile that sufficient solar absorptance requires small refractive index, preferably $n = 2$ or smaller, and that the absorptance product at all solar wavelengths should be 1.0 or more. Also low IR thermal emittance (or thermal absorptance) requires the absorption product to assume values near zero at a high reflectance of the base.

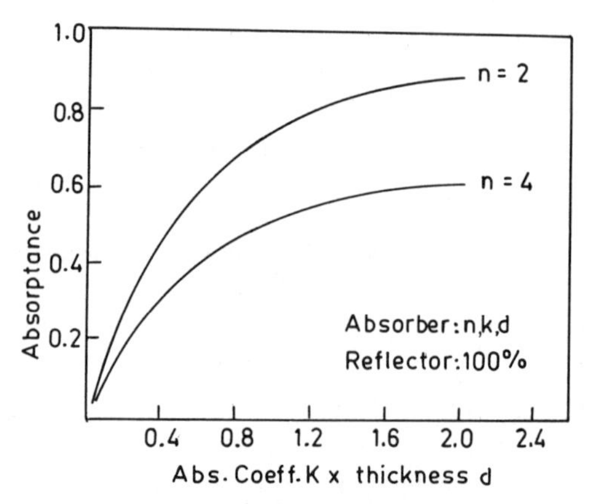

Fig. 5.3.1 Absorptance of an absorber–reflector tandem, as a function of absorption product *Kd* for different refractive indices. Adapted from Seraphin (1974) (Ref. 83).

In brief, a good tandem should meet the following requirements:

Solar Range:	Absorber	$Kd > 1$ and $n < 2$	
	Reflector	$0.6 < R < 1.0$	(5.3.2)
Thermal Range:	Absorber	$Kd \approx 0$ and n can be any value	
	Reflector	$R \approx 1.0$	(5.3.3)

Therefore for high solar absorptance, a material with as low refractive index as possible is required. This requirement is easily met out by a semiconductor, for which the fundamental absorption edge is located in the proper spectral range.

5.3.1 *Black Nickel*

In the United States in 1955, Tabor (64, 207–209), first described black nickel coatings (Ni–Zn–S alloy) over highly reflecting metal surface as solar selective surfaces. The black surfaces, which are produced by electrodeposition, have been in commercial use in solar water heaters in several countries for the last two decades. Recently McDonald (210) at NASA–Lewis Research Center, Mar et al. (211) at Honeywall Corporation, and several other authors (212–216) extensively studied black nickel coatings on stainless steel, mild steel, and nickel-plated mild steel. The black nickel coating is obtained by immersing the substrate as a cathode in an aqueous electrolytic bath containing per liter (208):

Nickel Sulfate	75 gm (105 gm)
Zinc Sulfate	28 gm (37 gm)
Ammonium Sulfate	24 gm (18 gm)
Ammonium Thiocynate	17 gm
Sodium Thiocynate	(15 gm)

The values shown in parentheses are taken by McDonald (210). The pH of the solution should be about 4 and a pure nickel anode should be used. The bath is operated at 30 °C. Electrolysis is carried on for 2–4 min at 2 mA/cm^2 current density. The exact time depends upon the nature of the substrate and the temperature of the bath. For better results the substrate should be coated with bright nickel which will enhance the thermal infrared reflectance and smoothness of the surface. The bright nickel coating on mild or galvanized steel sheet can be deposited by employing a bath containing per liter:

Nickel Sulfate	175 gm
Nickel Chloride	41 gm
Boric Acid	40 gm

The substrate is immersed as a cathode in a bath at 50–55°C. The pH value of the solution should be about 4 and current density during the deposition should be of the order of 3.5–4.5 A/cm^2. After removal from the plating, the panels are water rinsed, alcohol rinsed, and air dried.

Mar et al. (211) produced two-layer black nickel coatings with a broad reflectance minimum in the solar region by following the process developed by Tabor (208). The two-layer effect is accomplished by changing the current density during the deposition process. Tabor showed that there are two stable compositions which can be plated in this way, each having a different zinc content. Figure 5.3.2 shows the spectral reflectance of a two-layer black nickel on nickel-coated steel. Both the optical interference phenomenon and intrinsic absorption give rise to optical absorption in black nickel coatings. Similarly Fig. 5.3.3 shows the spectral reflectance of black nickel coating with different plating duration on a stainless-steel substrate and a comparison with idealized solar selective surface. It is seen that as the time of electroplating increases, there is a initial rapid increase in the absorptance with only a minimal accompanying increase in emittance. Then after approaching saturated value of absorptance, the emittance increases more rapidly. Therefore, a particular time of coating may be selected which gives the maximum solar absorptance with a minimum emittance. The optimum values of solar absorptance and thermal emittance were reported by McDonald and Curtis (210) for electroplated black nickel on stainless-steel selective surface as 0.84 and 0.18, respectively. The best results for black nickel on nickel substrate reported by Borzoni (212) are solar absorptance $\alpha_s = 0.96$ and thermal emittance $\varepsilon_T = 0.07$ but the author indicated poor humidity resistance. Certain improvements may be made by increasing the coating thickness but at the cost of thermal emittance. Sowell and Pettit (213)

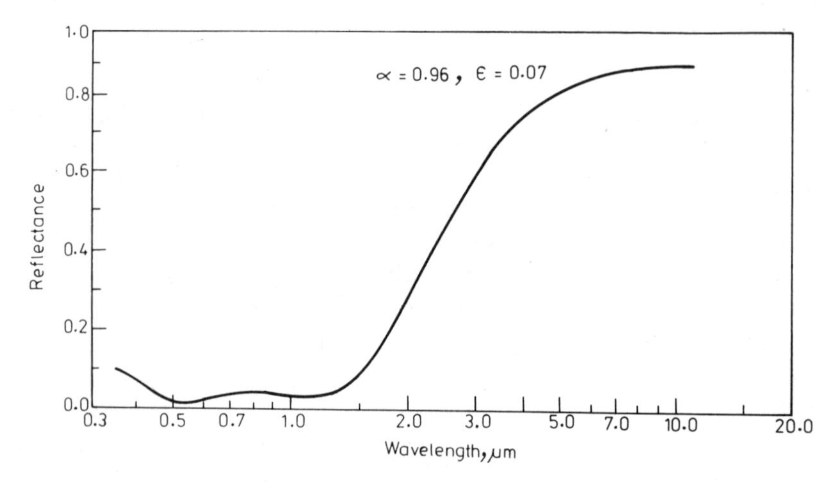

Fig. 5.3.2 Spectral reflectance of a two-layer black nickel on nickel coated steel. Adapted from Mar et al. (1976) (Ref. 211).

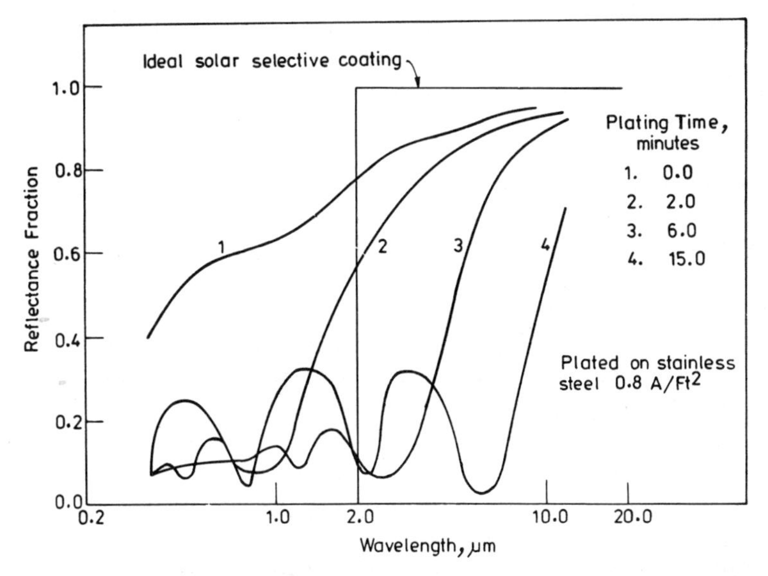

Fig. 5.3.3 Spectral reflectances of black nickel coatings with different plating times deposited on stainless-steel substrate. Adapted from McDonald and Curtis (1976) (Ref. 210).

have shown that black nickel coatings were unstable at 310°C in air and at lower temperature in the combined presence of heat and moisture.

5.3.2 *Black Chrome*

Recently, black chrome ($Cr–Cr_2O_3$ coatings) have been extensively investigated at NASA–Lewis Research Center by McDonald and coworkers (217–219) and at Honeywell Corporation by Mar and coworkers (220, 221). The plating time that would produce the optimum solar selectivity for black chrome plated over dull nickel or tin plated steel was investigated. The black chrome plating compound used is available as a proprietory mixture Chromonyx from the Harshaw Chemical Company (222) and E. I. DuPont Company (223). The nonproprietory formula has been presented in Ref. 224, but these coatings are unstable under UV irradiation in air at room temperature.

The test panels (217) were first plated with approximately 0.00005 in. of dull nickel (225) and then with bright nickel (218). The dull nickel was plated over with black chrome for varying length of time to produce a variation among the panels in thickness of black chrome. The panels were plated in the following sequence:

1. Cleaned by electrolytic alkaline chelating cleaner at 190°F and 70 to 80 A/ft^2. Two cycles, interspersed with acid, were used.
2. Dull nickel was plated from Harshaw Chemical Company Nusat at 40 A/ft^2 for 15 min to deposit approximately 0.0005 in. of nickel.

3. Black chrome was plated from Harshaw Chemical Company Chromonyx at 24 V and 200 A/ft^2 and different periods of time.
4. The panels were then rinsed in water, rinsed in alcohol and then dried in air.

McDonald (219) studied the technical feasibility of achieving optimum selectivity of black chrome plated on a commercially plated tin/steel substrate. The tin plated steel sheet was bright and uncorroded. The thickness of the tin was 0.0004 in. (0.001 cm). At 15 min plating time there was a heavy coating of black chrome over tin, but this was somewhat less heavy than the coating of black chrome produced by plating on dull nickel for 10 min. The values of solar absorptance and thermal emittance for various length of plating time are listed in Table 5.3.1.

The reflectance curves for black chrome as a function of plating times are shown in Fig. 5.3.4. It is obvious that both visible and IR reflectances decrease with increasing plating time. With increasing plating time and consequently increasing black chrome thickness, there is first a rapid increase in solar absorptance without a significant increase in thermal emittance until a nearly maximum value of absorptance is reached, after which the emittance rapidly increases with no appreciable increase in absorptance.

Figure 5.3.5 shows the solar absorptance and thermal emittance for a black chrome–nickel combination plotted against the black chrome thickness (226). The calculations of the balance between heat absorbed over the solar spectrum and heat radiated over the IR for a plate temperature of 250°F indicate that the optimum plating time for a particular current density (180 A/ft^2) was between 1 and 2 min. Lesser plating times would produce low thermal performance of solar collector due to low absorptance of the

Table 5.3.1 Solar Absorptance and thermal Emittance of Black Chrome Plated on Dull Nickel and Tin Plated Steel for Various Times

Black chrome on dull nickel	15 sec	30 sec	1 min	2 min	4 min	10 min
Solar absorptance	0.64	0.87	0.96	0.96	0.95	0.94
Emittance	0.04	0.06	0.10	0.12	0.17	0.34
Black chrome on tin/steel	13 sec	30 sec	1 min	2 min	4 min	15 min
Solar absorptance	0.80	0.84	0.96	0.97	0.96	0.94
Emittance	0.08	0.07	0.06	0.06	0.07	0.18

Source: Adapted from McDonald (1977), (Ref. 219).

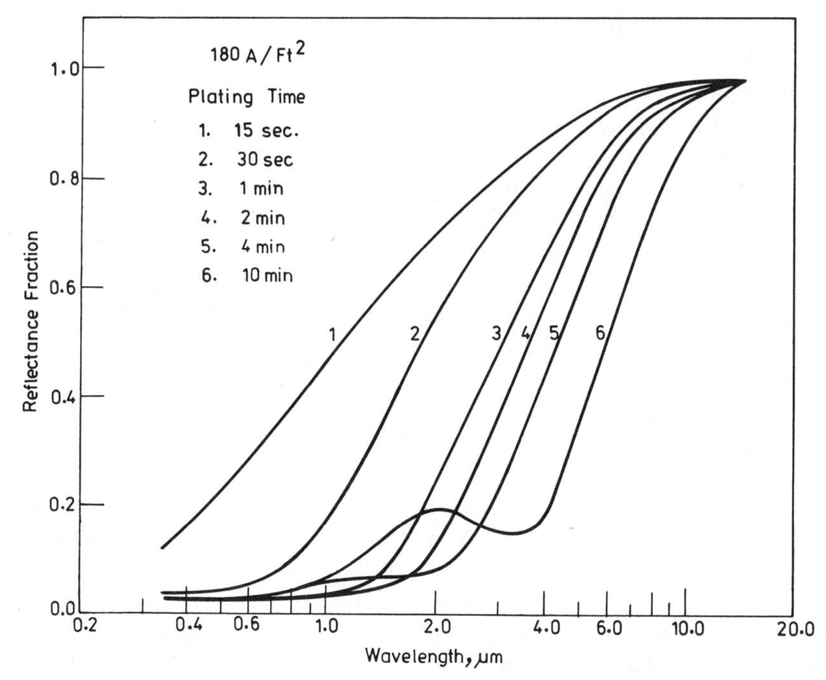

Fig. 5.3.4 Spectral reflectances of black chrome coatings deposited for different plating times on nickel plated steel. Adapted from McDonald and Curtis (1977) (Ref. 219).

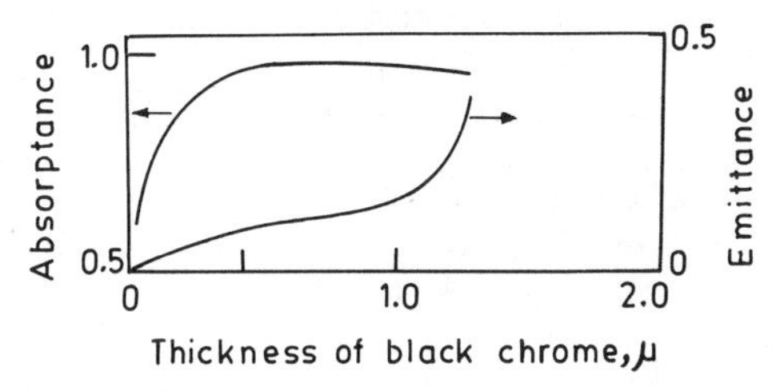

Fig. 5.3.5 Solar absorptance and thermal emittance of black chrome versus coating thickness. Adapted from McDonald et al. (1976) (Ref. 226).

plate. Greater plating time would produce low thermal performance of a solar collector due to a high loss resulting from high plate emittance.

Pettit and Sowell (227) also prepared black chrome solar selective surfaces by electrodeposition of Chromonyx Harshaw solution onto bright nickel, sulfamate nickel, Watts nickel (224), Zodiac nickel (222) and Nusat nickel (222) substrate. Solar absorptance and thermal emittance were studied as a function of electroplating time and substrate surface roughness. Both solar absorptance and thermal emittance increase with plating time, but for large plating time there is no appreciable change in solar absorptance although thermal emittance keeps on increasing. It was also noticed that by increasing surface roughness the solar absorptance increases while emittance remains unchanged.

Mar et al. (211, 220, 221) have also studied in detail the solar selective properties of black chrome coatings [prepared from DuPont solution (223) and Harshaw (228) solution] on nickel plated, bare and galvanized steel, and copper substrates. The total reflectance properties were studied for chrome plated samples subjected to MIL-STD-810 B humidity test. After 8 days of exposure, the sample showed no sign of deterioration. The solar absorptance and thermal emittance remained at 0.95 and 0.11, respectively. Figure 5.3.6 shows the reflectance of black chrome on nickel plated steel for a 500 mA-min/cm^2 time and current density product. Figure 5.3.7 shows the reflectance of black chrome on nickel plated steel after 8 days of a cycling humidity test.

Harshaw black chrome (228) was also plated directly onto steel substrate. After cleaning the steel in a solution of hydrochloric acid and water the black chrome was plated. The coatings were matte black and had a sooty appearance. After 24 hr under MIL-STD-810B humidity test, numerous small rust spots covered the sample surface. To improve the humidity resistance, black chrome was plated onto bare steel, then covered with an organic overcoat. The spectral reflectance of the final coating is shown in Fig. 5.3.8. The solar absorptance was 0.94 and emittance was 0.20.

Black chrome was deposited on galvanized-steel sheet (211, 220). The substrate was first placed in a dilute hydrochloric–chromic acid solution for about 1 min. This procedure produces a uniform brown film that covers the substrate surface. This film was removed by immersing the sample in a

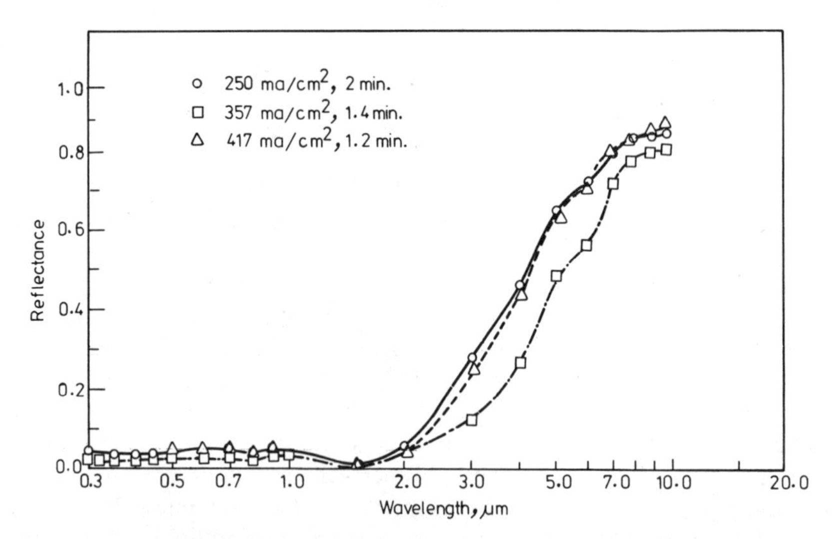

Fig. 5.3.6 Spectral reflectance of black chrome on nickel plated steel for a time and current density product of 500 mA-min/cm^2 for different time and current density. Adapted from Mar et al. (1976) (Ref. 211).

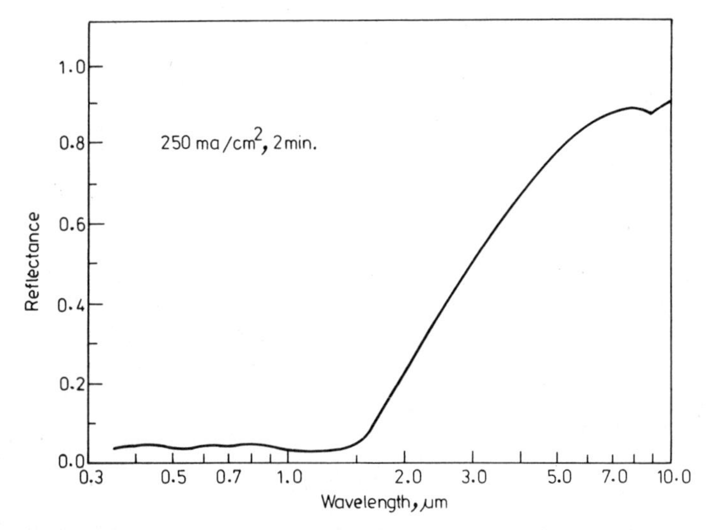

Fig. 5.3.7 Spectral reflectance of black chrome on nickel coated steel after eight days humidity exposure. Adapted from Mar et al. (1976) (Ref. 211).

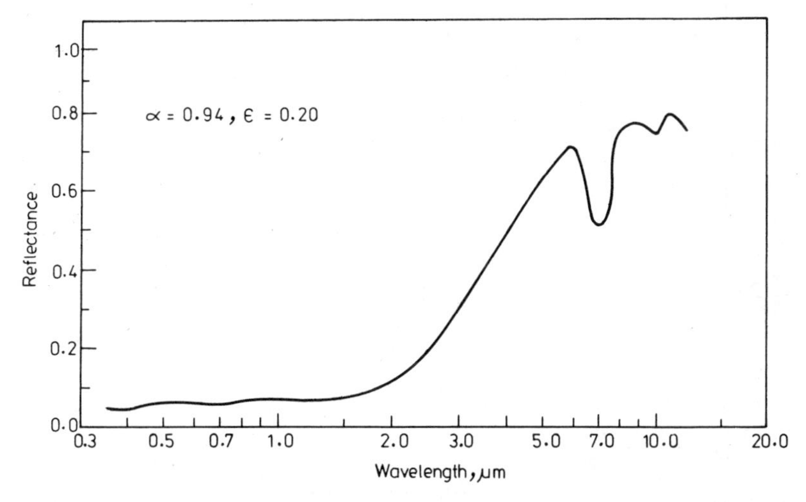

Fig. 5.3.8 Spectral reflectance of black chrome on steel with organic overcoat. Adapted from Mar et al. (1977) (Ref. 221).

solution of chromic acid and water. The black chrome coating was plated by the usual technique. The coatings had average solar absorptance of 0.95 to 0.96 and average thermal emittance of 0.15. Figures 5.3.9 and 5.3.10 show the reflectance versus wavelength plots for various plating times and current densities for black chrome on galvanised-steel sheet. The visual inspection of the sample after 24 hr exposure under MIL-STD-810B humidity test showed sufficient degradation.

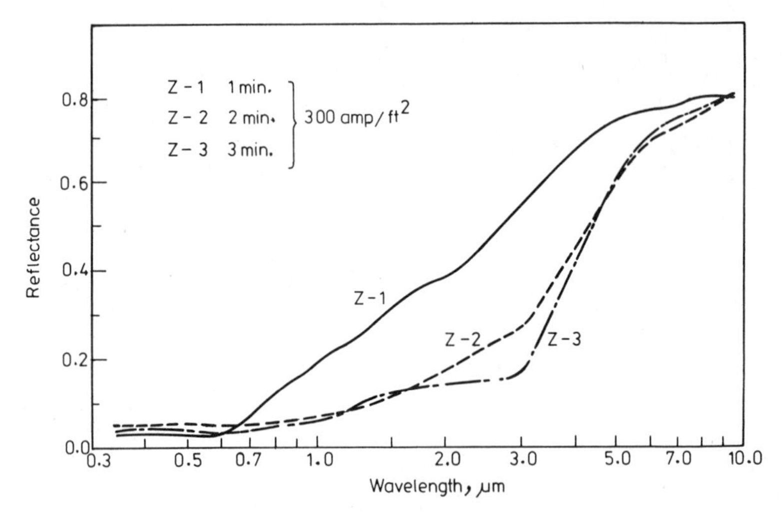

Fig. 5.3.9 Effiect of plating time on the spectral reflectance of black chrome, deposited on galvanized steel sheet. Adapted from Mar et al. (1976) (Ref. 211).

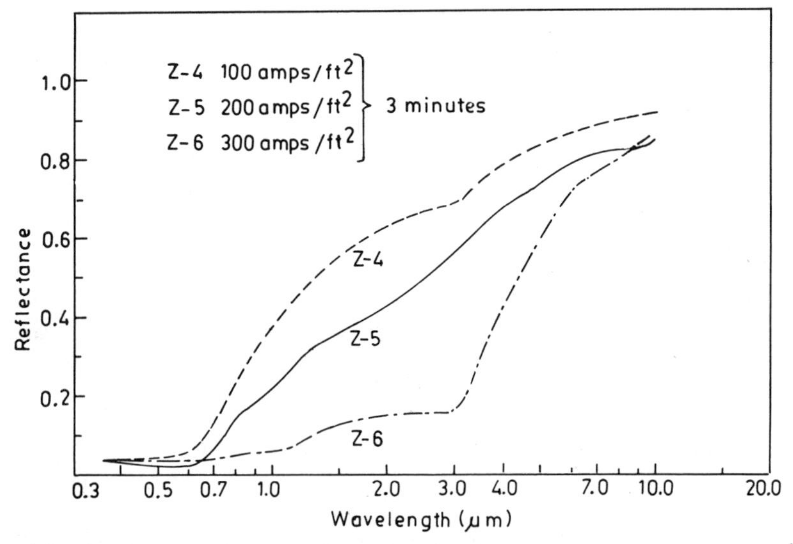

Fig. 5.3.10 Effect of current density on the spectral reflectance of black chrome, deposited on galvanized-steel sheet. Adapted from Mar (1976) (Ref. 211).

Black chrome coatings were applied on copper substrate using Harshaw black chrome solution (211). The coatings have average solar absorptance of 0.95 and thermal emittance of 0.14 and good humidity resistance. Black chrome on copper shows good selectivity and humidity resistance but is not feasible because of the high cost of copper substrate. Figure 5.3.11 shows reflectance versus wavelength plot for black chrome on copper sheet selective surface.

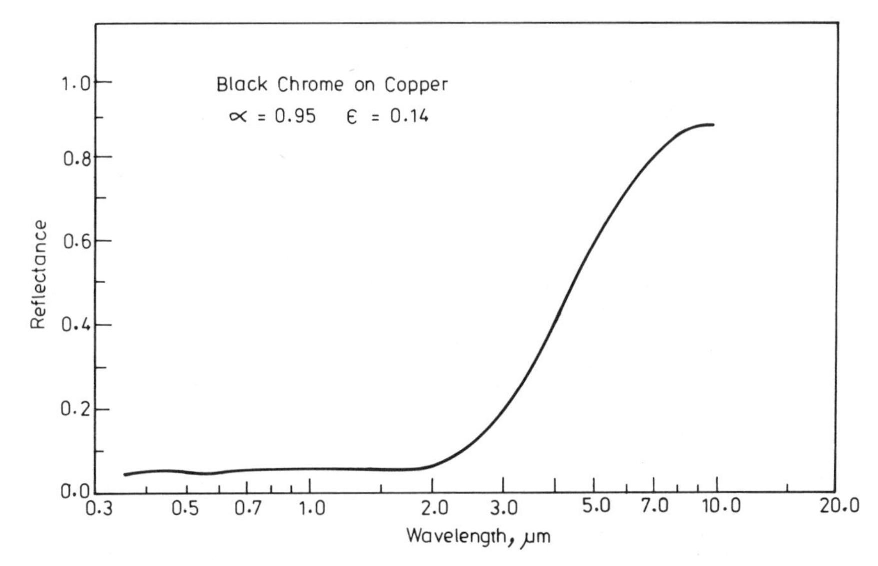

Fig. 5.3.11 Spectral reflectance of black chrome deposited on copper sheet. Adapted from Mar et al. (1976) (Ref. 211).

In recent years several authors (229–231) have investigated surface microstructures and correlated them with optical properties of black chrome coatings produced by electrodeposition and vacuum evaporation. Lampert (85) found that black chrome consisted of a very fine metallic distribution of particles of chromium, possibly suspended within a matrix of an oxide of chromium. The selective properties of various coatings investigated by several authors are listed in the Table 5.3.2.

5.3.3 *Black Copper*

Black copper oxide coating on copper or aluminium is the most commonly used selective surface and is extensively commercialized in solar collector industries. Several authors (207, 209, 210, 232–238) investigated solar selective properties of CuO coatings prepared by various processes, e.g., spraying, chemical conversion, chemical brightening and etching, etc. Hottel and Unger (235) produced copper oxide layer by spraying a dilute solution of cupric nitrate onto a heated aluminium sheet. Under these conditions a light green layer was formed on the aluminium substrate, which, upon heating above 170°C, was converted into the black cupric oxide layer. Hottel and Unger (235) studied the dependence of absorptance and emittance on various preparative parameters like concentration of nitrate in spray, spray particle size, spray rate, drying plate temperature, baking temperature, amount of deposit per unit area, etc. The solar absorptivities of these coatings were dependent on the mass per unit area of the film. The thermal emittance was found to be decreased by decreasing solution concentration and decreas-

Table 5.3.2 Solar Absorptance α, Thermal Emittance ε, and Selectivity α/ε of Electrodeposited Black Chrome Coating on Various Substrates

No.	Author/ Ref. No.	Substrate	Solar Absorp-tance, α	Thermal Emittance, ε	α/ε
1.	McDonald (218, 219)	Dull nickel plated steel	0.96	0.10	9.6
2.	McDonald (218,219)	Tin plated steel	0.96	0.06	16.0
3.	Mar. et al. (211)	Nickel plated steel	0.95	0.11	8.6
4.	Mar. et al. (211)	Steel	0.94	0.20	4.7
5.	Mar. et al. (211)	Galvanized steel	0.95–0.96	0.15	6.3
6.	Mar. et al. (211)	Copper sheet	0.95	0.14	6.7
7.	Borzoni (212)	Nickel	0.95	0.11	8.6
8.	Keeling et al. (232)	Aluminium foil	0.964	0.023	42.0
9.	McDonald et al. (233)	High zinc die-cast and Harshaw Neostratm AFA plated zinc	0.95	0.07	13.2
10.	McDonald et al. (233)	Zincated aluminium	0.95	0.09	10.5
11.	Pettit and Sowell (227)	Bright nickel	0.92	0.11	8.3
12.	Pettit and Sowell (227)	Nusat nickel	0.95	0.11	8.6
13.	Lampert (85)	12.7 μm Thick Ni coated steel	0.95	0.07	13.7
14.	Lampert (85)	12.7 μm thick Ni coated copper	0.95	0.078	12.3
15.	Lampert (85)	Steel	0.92	0.07	13.1
16.	Lampert (85)	Copper	0.94	0.51	18.5
17.	Harding (230)	Deposited in vacuum on polished copper substrate	0.88	0.05	17.2

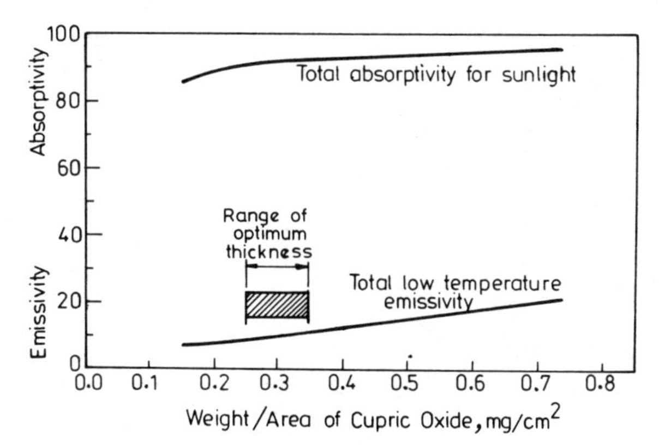

Fig. 5.3.12 The dependence of solar absorptance and thermal emittance on CuO coating thickness. Adapted from Hottel and Unger (1959) (Ref. 235).

ing spray droplet size, and was lowest after baking at 350°C. The optimum values of solar absorptance and thermal emittance were 0.93 and 0.11, respectively. The dependence of solar absorptance and thermal emittance of CuO coatings under optimum conditions is shown in Fig. 5.3.12.

In 1956 Tabor (207, 209, 232) developed a method of applying black copper oxide selective surface on aluminium and copper substrate by electroplating. The aluminium base was first covered with an oxide layer by anodizing. For this purpose the aluminium sheet was immersed as a cathode in an aqueous solution containing 3% by volume of sulfuric acid and 3% by volume phosphoric acid with carbon as an anode. An electric current of 6 mA/cm^2 was passed through the solution for 20 to 30 sec. After rinsing, the aluminium sheet was immersed for 15 min at 85–90°C in an aqueous solution containing 25 g/liter of copper nitrate, 3 g/liter of nitric acid, and 15 g/liter of potassium permanganate. After this process the aluminium sheet was dried and heated to about 450°C for few hours (207, 209). For black copper oxide coating a copper substrate was buffed and degreased and was treated for various times (between 3 and 13 min) in a bath at 140–145°C. The bath consists of 16 oz sodium Hydroxide, and 8 oz sodium chloride per gal of water. This formulation was followed by Tabor (207) and Salem and Daniels (234) for producing black copper oxide selective surfaces. The values of a solar absorptance of 0.79 and a thermal emittance of 0.05 were reported for a sample treated for 3 min. For an 8 min treatment time, the solar absorptance and thermal emittance were 0.89 and 0.17, respectively. Thermal emittance increases with treatment time. Mar et al. at Honeywell Incorporated (211) also prepared copper oxide coatings on copper by the above process. A solar absorptance of 0.90 and thermal emittance of 0.15 were about the best that could be achieved with copper oxide coatings. An increase in solar absorptance was observed when CuO was deposited on a rough surface, but the coating was easily damaged. In the above oxidation process, Christie (236) suggested that the chemical

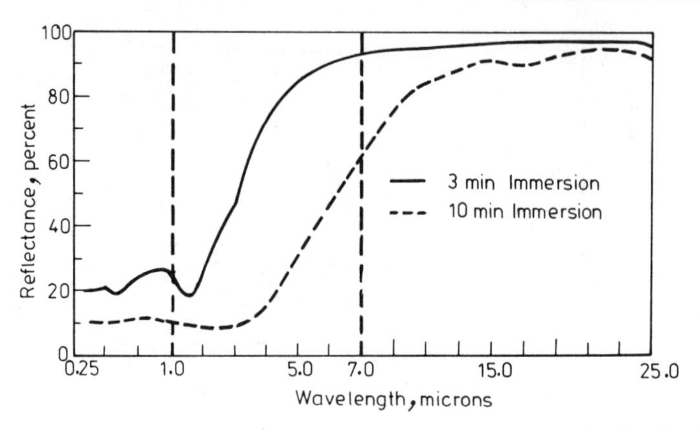

Fig. 5.3.13 Spectral reflectance of CuO copper black on copper deposited by 3 and 10 min immersion time. Adapted from Christie (1970) (Ref. 236).

composition of the surface consist of predominantly cupric oxide with cuprous oxide and traces of other compounds formed by secondary radiations. The following reactions most likely occur during the oxidation process:

$$\begin{aligned} 2\,Cu + 2\,NaClO_2 &\rightarrow Cu_2O + NaClO_3 + NaCl \\ Cu_2O + 2\,NaClO_2 &\rightarrow 2CuO + NaClO_3 + NaCl \end{aligned} \tag{5.3.4}$$

The spectral reflectance of copper black is shown in Fig. 5.3.13. The immersion in the alkaline oxidation bath for 3 min gives solar absorptance $\alpha_s = 0.79$ and thermal emittance ε at 150°F = 0.05, whereas immersion for 10 min increases α_s to 0.90 and ε at 150°F to 0.20.

Mattox and Sowell (242) and McDonald and Curtis (210) produced black copper oxide coatings by chemically blackening copper using Ebonol C (trade name, Enthone Inc., USA). The as-treated surface has a velvet appearance owing to dendritic structure (Fig. 5.3.14) and is normally burnished after treatment. X-ray diffraction studies show that the surface is composed mostly of CuO with some Cu_2O. McDonald and Curtis reported that for increasing time of oxidation, the general appearance of the panels varied from the untreated copper to a very intense jet black. The results of spectral reflectance of black copper panels oxidized for various times up to 4 min are shown in Fig. 5.3.15. At the longer time of oxidation there is a considerable decrease in infrared reflectance. The optimum values of solar absorptance $\alpha_s = 0.91$ and thermal emittance $\varepsilon = 0.05$ were found for a 1 min coating formation time. Figure 5.3.16 shows the effect of bath temperature on spectral reflectance of black copper coatings on copper (211). The high temperature stability of the coatings is limited due to further oxidation producing defoliation (237), and corrosion stability (238) is also not very good. Under high humidity (100% relative humidity) and 38°C condition deterioration was observed after 500 hr.

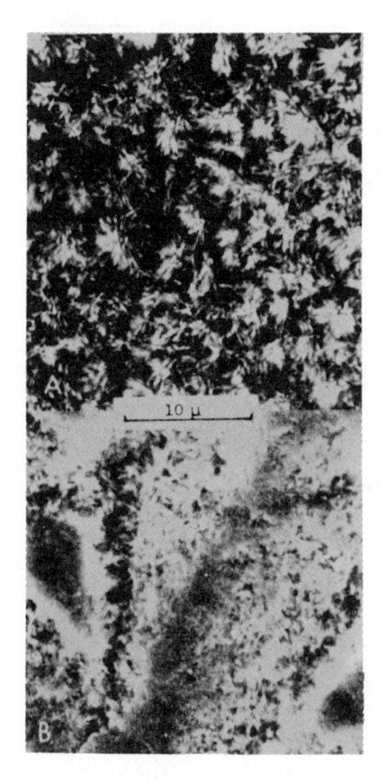

Fig. 5.3.14 Surface texture of Ebonol C treated copper (A) as treated, (B) after burnishing. Adapted from Mattox and Sowell (1974) (Ref. 242).

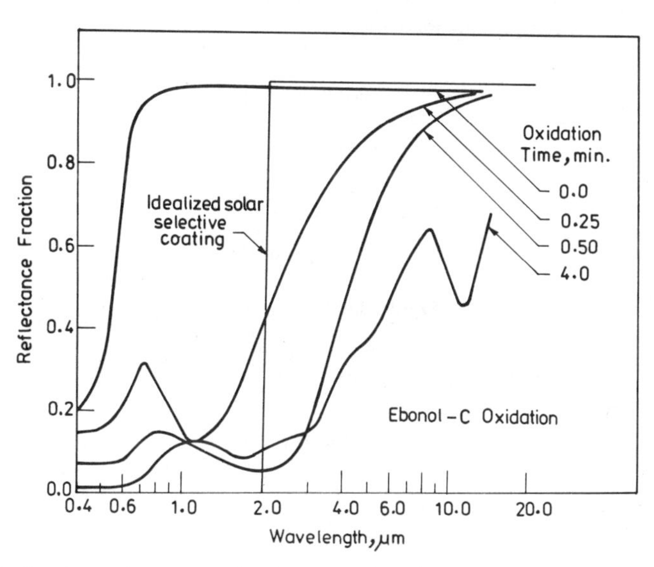

Fig. 5.3.15 Spectral reflectance of black copper deposited under different oxidation times adapted from McDonald and Curtis (1975) (Ref. 255).

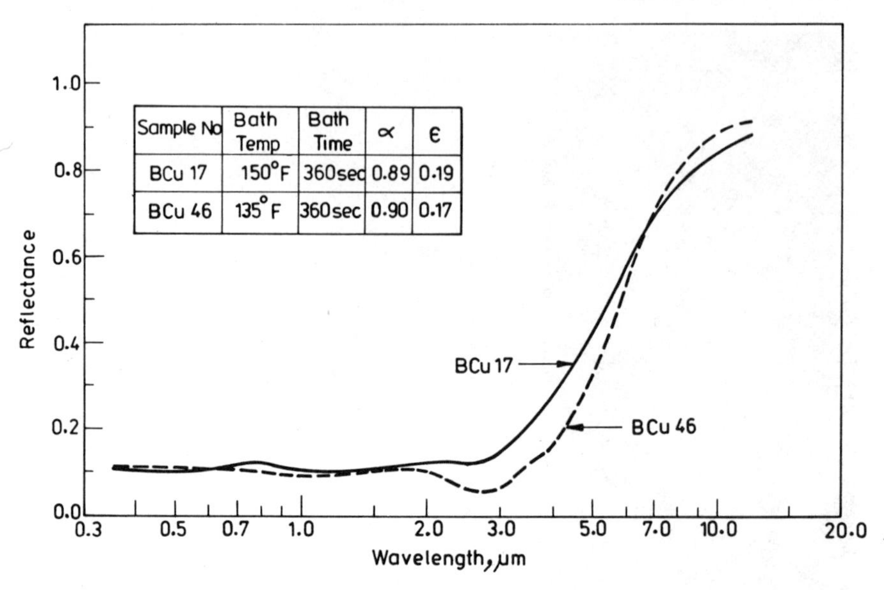

Fig. 5.3.16 The effect of bath temperature on spectral reflectance of black copper. Adapted from Mar et al. (1976) (Ref. 211).

Reid and Wilson (240) found that black copper could not withstand 350°F for 30 days. Exposure of black copper to the MIL-STD-810B humidity test for 24 hr resulted in a complete coating removal. Some efforts are underway at International Copper Research Association by Lindstram et al. (239) to improve the durability by the application of acrylic layers containing a copper resistance inhibitor.

Table 5.3.3 Composition of Selected Brightening and Etching Solutions

Solution	Composition	Temperature
A	75% H_3PO_4, 25% H_2SO_4 Chemical brightening solution	100°C
B	49%, H_2O, 34% HCl, 15% HNO_3, 2% HF etchant	23°C
C	Solution B + 0.5% cupric nitrate	23°C
D	20% NaOH, 80% H_2O alkaline etchant	66°C

Source: Adapted from Driver and McCormik (1978) (Ref. 241).

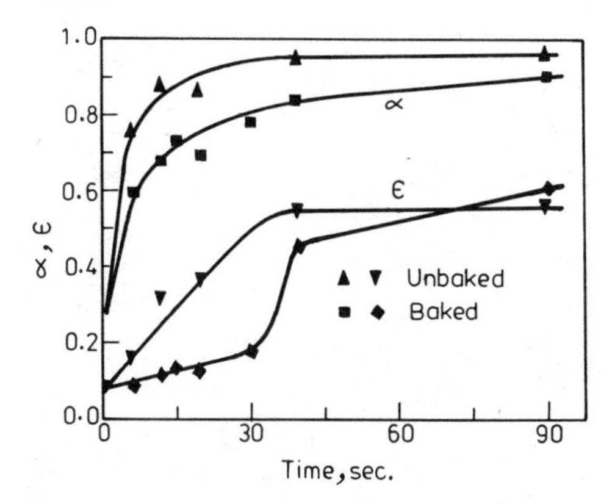

Fig. 5.3.17 Dependence of heat treatment time on absorptance and emittance. Adapted from Driver and McCormic (1978) (Ref. 241).

Driver and McCormic (241) followed an alternative method of producing CuO films on an aluminium surface by method of chemical brightening and etching followed by baking. The material used in this study was a commercial aluminium–copper alloy having a nominal composition of 4.5% Cu, 1.5% Mo, 0.7% Mn, 0.5% Fe, and 0.5% Si. The substrates were mechanically polished and degreased prior to chemical treatment. Four different chemical brightening and etching solutions were investigated whose compositions are listed in Table 5.3.3. The samples were immersed in the solutions for varying length of time, followed by washing in distilled water and drying in warm air. Baking was carried out at 350°C for 90 min. The effect of immersion time on the solar absorptance and emittance is shown in Fig. 5.3.17 for samples prepared using solution A. Both the absorptance and emittance increased with increasing treatment time, reaching values of $\alpha_s = 0.95$ and $\varepsilon_T = 0.55$, respectively, for an immersion time of 40 sec. The baking of the surface at 350°C for 90 min was found to result in a decrease to both α and ε as shown in Fig. 5.3.17. The surface morphology of the coatings using solution B was investigated (241) using a scanning electron microscope. Unbaked coatings formed after 6 and 12 sec treatments are shown in Fig. 5.3.18(a) and 5.3.18(b), respectively. The coatings consist of a sponge-like aggregate of particles arranged in a cellular array. The diameter of the cells is approximately 1 μm and appears to be constant independent of treatment time. The size of particles in the coating is of the order of 0.1 μm. It was noticed that baking time did not alter the cellular structure of the coating or the cell size [Fig. 5.3.18(b)]. At higher magnification, it appeared that particle scattering accompanied the conversion of Cu to CuO. The cellular morphology suggests the possibility of a cavity absorption mechanism since the cell size is on the order of 1 μm. The increased coating weight required for high absorption has resulted in higher emittance than that obtained by Hottel and Unger (235) at comparable absorptance, in particular $\varepsilon_T > 0.16$ for $\alpha_s > 0.90$.

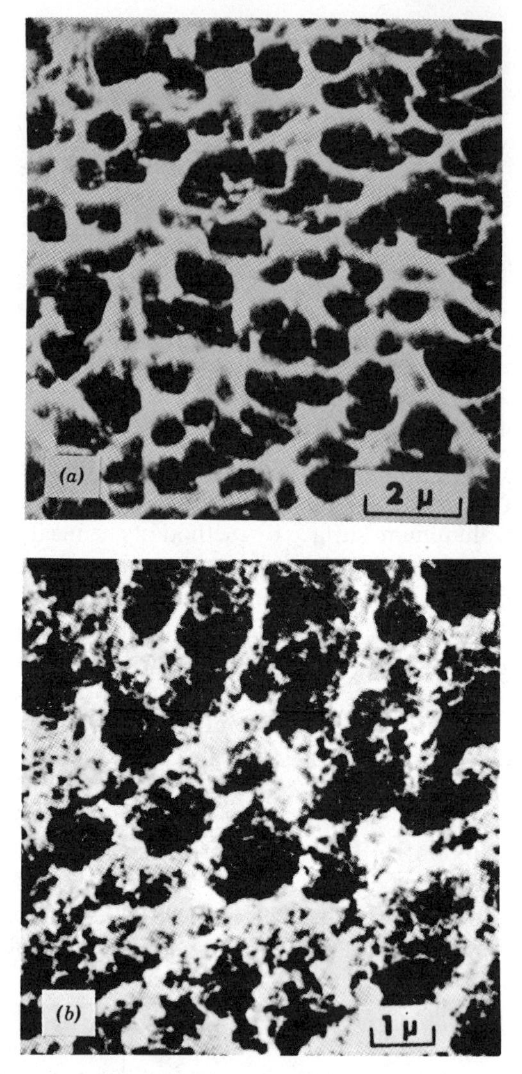

Fig. 5.3.18 Surface texture of unbaked coating treated for (a) 6 sec and (b) 12 sec in solution. Adapted from Driver and McCormic (1978) (Ref. 241).

5.3.4 *Black Iron*

Several authors (211, 221, 236, 224, 243–247) have reported a number of thermal and chemical processes for producing an iron oxide coating which is easy and inexpensive. Mar et al. (211) have made extensive studies on the Ebonol S process of Enthone, Inc (248) for black iron coatings. Lin et al. (221) studied in detail the various possibilities of increasing the solar absorptance of these coatings. Chirstie (236) also produced these coatings from a commercial bath named Black Coat MR (Holland) containing a strong alkaline solution. After immersion of steel substrate for seconds at boiling temperature of 140°C, a smooth, uniform, and dark coating is achieved. In the Ebonol S process the iron oxide coating is produced by immersing a steel substrate into a caustic Ebonol S solution, which is heated just under the boiling point. Auger analysis of the Ebonol coatings showed that the coating is $\approx 48\%$. Iron and 52% oxygen. The oxide formed is Fe_3O_4. A cold rolled, low-carbon steel sheet is used, which has high corrosion resistance. The steel surface is prepared by immersion in dilute hydrochloric acid for approximately 10 sec. The optimum values of solar absorptance and thermal emittance are 0.85 and 0.10, respectively. Figure 5.3.19 shows the effect of immersion time on spectral reflectance. Mar et al. (211) studied spectral reflectance by varying different preparative parameters like immersion time, bath temperature, and solution concentration. These iron oxide coatings have shown reasonably good humidity resistance and survived for 4 to 5 days under MIL-STD-810B humidity test with only minor signs of degradation.

Mar et al. (211) also investigated two alternative methods for producing iron oxide selective coatings. One method was simply to heat the sample in air. The spectral reflectance is shown in Fig. 5.3.20. These coatings did not show the same degree of humidity resistance as did the Ebonol S coatings. The plate developed a moderate number of small rust spots after 1 day. The second method was a immersion process at room temperature in Presto Black (247).

It was suggested that the solar absorptance of iron oxide coating could be

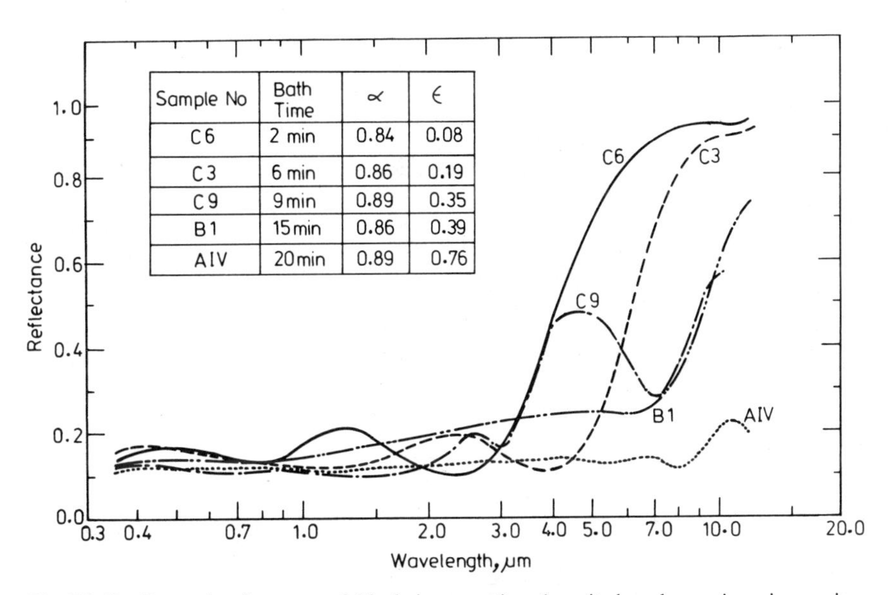

Fig. 5.3.19 Spectral reflectance of black iron coating deposited under various immersion times. Adapted from Mar et al. (1976) (Ref. 211).

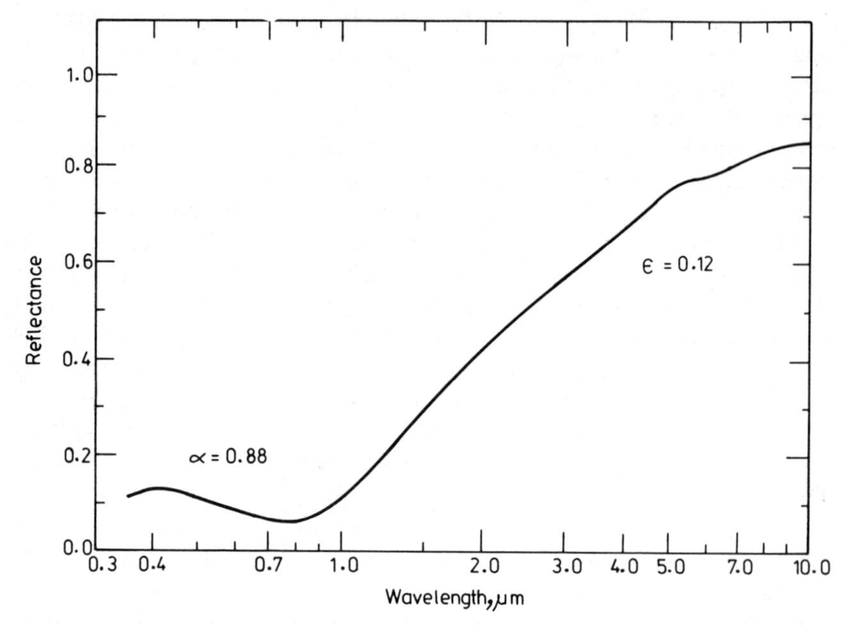

Fig. 5.3.20 Spectral reflectance of iron oxide selective surfaces produced by heating steel in air. Adapted from Mar et al. (1976) (Ref. 211).

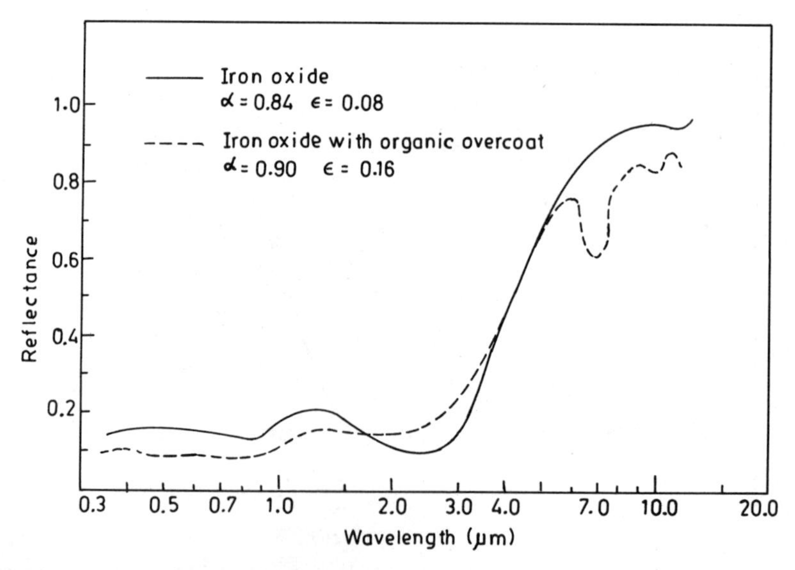

Fig. 5.3.21 Spectral reflectance of as-grown and organic-overcoated iron oxide selective surfaces. Adapted from Mar (1976) (Ref. 211).

increased by overcoating by organic polymer which acts as an antireflection coating. However it simultaneously increases the thermal emittance because of absorption bands in the infrared region.

Mar et al. (211) overcoated iron oxide samples by three organic overcoats; namely, Dow Corning XR-6–2205 (249), Exxon Vistalon 6505—an ethylene–propylene–diene material (EPDM) (250), and Exxon Vistalon 606—an ethylene–propylene material (EPM) (250). The EPDM and EPM overcoated samples showed reasonably good humidity resistance. Figure 5.3.21 shows the spectral reflectance of iron oxide coating before and after overcoat with EPDM. An increase of 0.06 was observed in solar absorptance after overcoating. Lin et al. (221) have made some efforts to improve solar absorptance of iron oxide coatings by overcoating a thin layer of black nickel over black iron. An improvement of 0.03 in solar absorptance was observed without any change in emittance value.

5.3.5 *Cobalt Oxide*

Vander Leij (251, 252) prepared cobalt oxide selective surfaces on bright nickel plated steel substrates by electroplating technique. The electrolyte bath was based on the recipe of Srivastava and Kumar (253). Srivastava and Kumar (253) carried out extensive study of the porosity, microstructure, and adhesion of cobalt electrodeposits under different electroplating conditions. They suggested various bath compositions for cobalt plating, but Vander Leij used the plating bath containing 450 g/liter $CoSO_4 \cdot 7H_2O$, 45 g/liter $CoCl_2 \cdot 6H_2O$ and 40 g/liter H_3BO_3. The bath temperature was 400°C and the current density 4 A/dm^2. The pH was made 4 by adding a

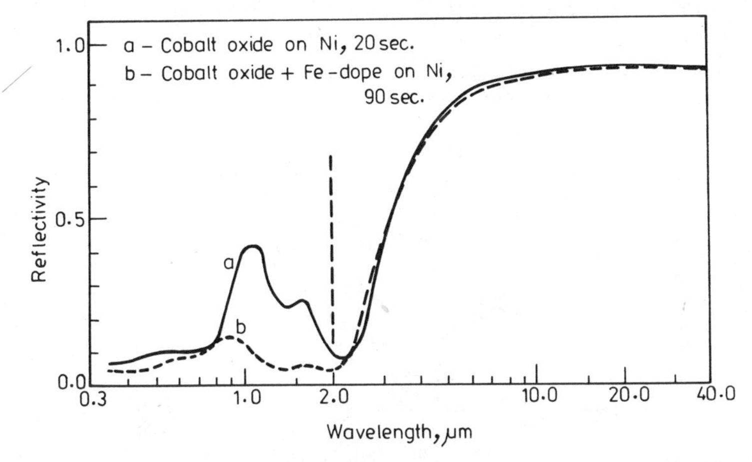

Fig. 5.3.22 Spectral reflectance of cobalt oxide on nickel plated steel produced by (a) oxidized cobalt electroplated in baths consists of 450 g/liter $CoSO_4 \cdot 7H_2O$, 45 g/liter $CoCl_2 \cdot 6H_2O$, and 40 g/liter H_3BO_3 at 40°C; (b) previous bath plus $Fe_2(SO_4)_3$. Adapted from Vander Leij (1978) (Ref. 251).

small amount of Co $(OH)_2$ and plating time was 20 seconds. After plating, the cobalt was converted into cobalt oxide by heat treatment at 400°C for two hours. The oxide formed this way was nearly cobalt oxide with a small excess of oxygen. The spectral reflectivity is shown in Fig. 5.3.22. A high reflectivity at about 1.2 μm was observed. Vander Leij suggested that a doping of Fe^{3+} into CoO reduced the reflectivity peak at 1.2 μm in spectral reflectance curve considerably. He added a small amount of $Fe_2(SO_4)_3$ to the cobalt sulfate bath. The solar absorptance of 0.87 and 0.92 and thermal emittance of 0.07 and 0.08 were observed in CoO and Fe doped CoO selective coatings, respectively, deposited on nickel plated steel substrate. A slight enhancement in solar absorptance and thermal emittance was noticed on Fe doping.

5.3.6 *Tungsten Oxide*

Vander Leij (251) produced tungsten oxide solar selective surfaces by RF reactive sputtering technique. The films were deposited onto polished nickel or nickel plated steel substrate. The sputtering process was carried out in a gas atmosphere of 97% argon and 3% oxygen with a vacuum pressure of 8.10^{-3} torr during 8 min. The substrate temperature was varied upto 300°C. The tungsten oxide formed was nonstiochiometric WO_3. An excess of tungsten was observed. Figure 5.3.23 shows the spectral reflectance of RF sputtered tungsten oxide coating and Al_2O_3 overcoated oxide coating. The high reflectance of WO_3 in solar region can be reduced by overcoating an antireflection coating onto it. The surface of the coatings were quite uniform. Large-area coatings produced by RF sputtering are described by Thornton (254). The solar absorptance of 0.83 and 0.93 and thermal emittance of 0.07 and 0.09 were found in WO_3 and Al_2O_3 overcoated WO_3 coatings

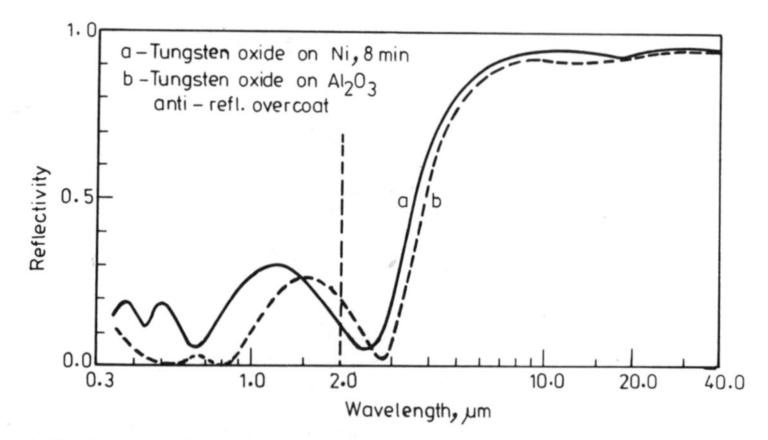

Fig. 5.3.23 Spectral reflectance of tungsten oxide on polished nickel produced by (a) RF reactive sputtering in 97% Ar/3% atmosphere at 8×10^{-3} torr vacuum, (b) overcoated with Al_2O_3 antireflection coating. Adapted from Vander Leij (1978) (Ref. 251).

deposited on nickel plated steel samples. Vander Leij observed that tungsten oxide selective surfaces are unstable at elevated temperatures. They proved somewhat volatile after a short heating at 400°C. However, after overcoating tungsten oxide by Al_2O_3, the films were quite stable and after a heat treatment of 2 hr at 400°C, the solar absorptance changed from 0.93 to 0.95 and thermal emittance from 0.09 to 0.12.

5.4 Conversion Coatings

Chemically converted coatings generally used for decorative purposes are easy to produce, low in cost, and have wide availability. Several researchers investigated solar selective properties of colored stainless steel, copper oxide, copper sulfide, chromate, and chloride conversion of zinc and ALCOA 655 coatings. The chemical conversions were made by the same standard solutions available from Harshaw chemical company. In the next section we discuss the properties and preparation of various chemically converted solar selective surfaces.

5.4.1 *Copper Sulfide*

Copper sulfide is a direct gap semiconductor having 1.8 eV band gap. This coating can be prepared by treating copper surface in a bath containing ammonium sulfide solution. Mattox and Sowell (242) produced copper sulfide selective surfaces by this method. Figure 5.4.1 shows the spectral reflectance and transmittance of chemically converted Cu_2S films. The solar absorptance of the coating was only 0.79 since it does not absorb in the near IR region of the spectrum. For the IR transmittance, the Cu_2S film was formed by completely sulfiding a thin (1 μm) copper film and then grinding the film to a powder and pressing in a KBr pellet.

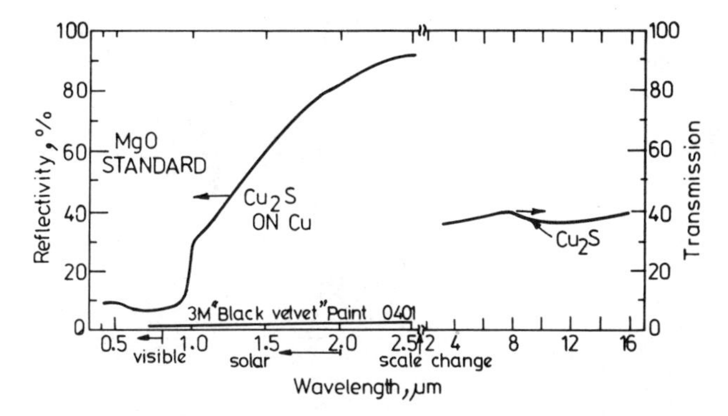

Fig. 5.4.1 Spectral reflectance and transmittance of Cu_2S coating produced by chemical conversion of copper sulfate. Adapted from Mattox and Sowell (1974) (Ref. 242).

5.4.2 *Black Zinc Coatings*

McDonald and Curtis (255) at NASA–Lewis Research Center investigated solar selective properties of electroplated and chemically converted zinc coatings. Converted electroplated zinc was produced by chromate and chloride conversion of the standard Neostar^{TM} chromate black solution, a

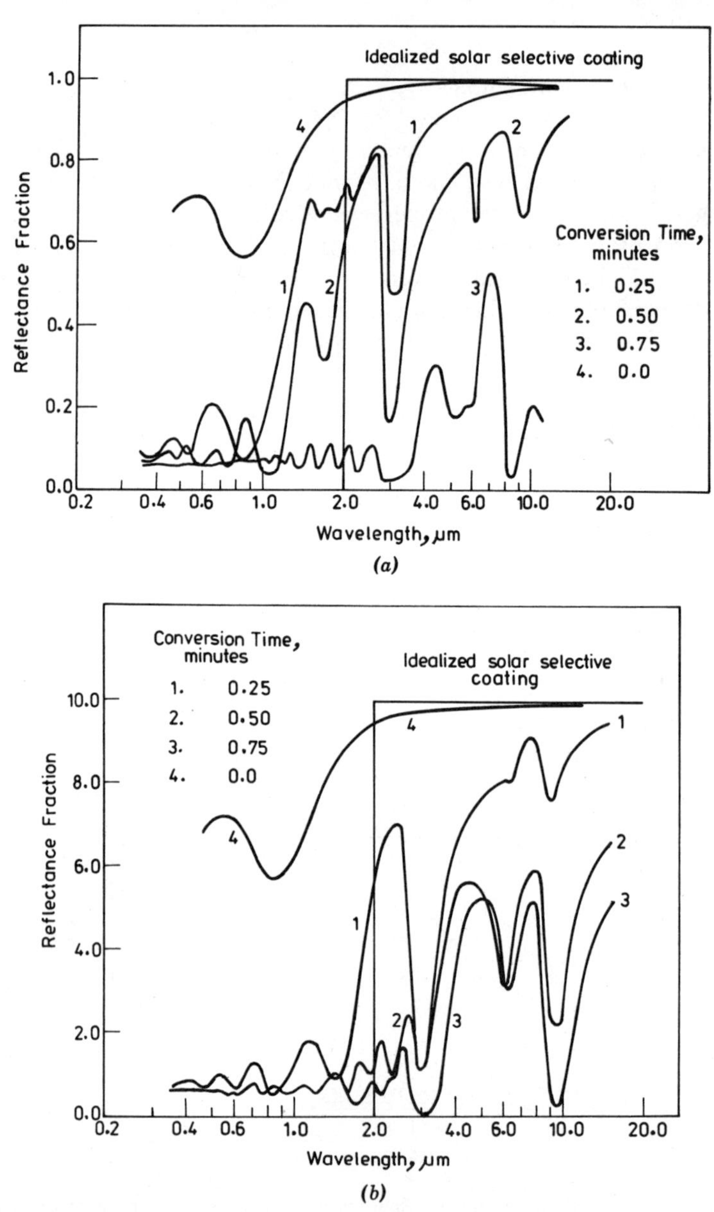

Fig. 5.4.2 Spectral reflectances of black zinc produced by (a) 50% (b) 75% (c) 100% concentration in chromate conversion. Adapted from McDonald and Curtis (1975) (Ref. 255).

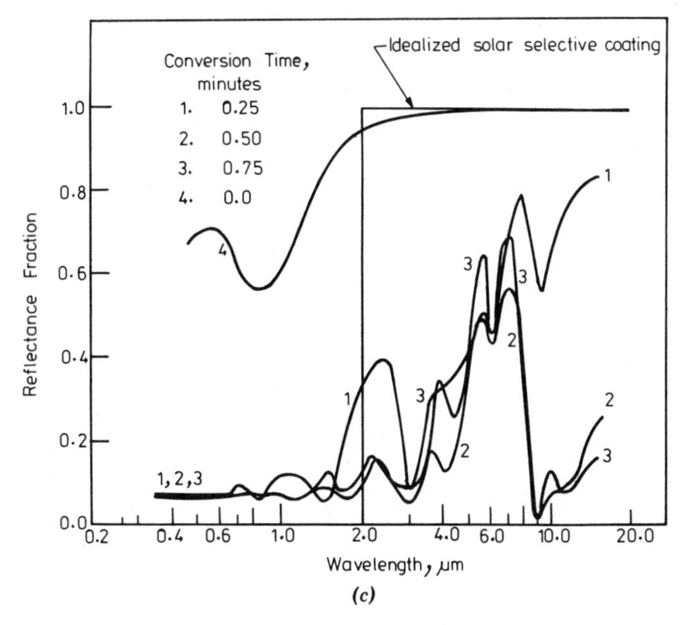

(c)

Fig. 5.4.2 (*Continued*)

proprietory process of Harshaw Chemicals. A nonproprietory method of preparation is described by Encheva (256, 257) and Vander Leij (251) for blackening zinc anodically in a solution with NaOH and $NaNO_3$ with alternating current (A.C.)

The samples were prepared (255) by electroplating zinc compound (available as a proprietory material from Harshaw chemicals) and then dipping them in solution of conversion compound with agitation for some time. The solar selective properties were studied as a function of conversion treatment time. The chromate conversions were carried out at 50, 75, and 100% concentration of that normally used for Neostar™ black chromate. For increasing times, the appearance of samples varied from red through violet to jet black. The spectral reflectances of electroplated and converted zinc as a function of wavelength are shown in Figs. 5.4.2(a), 5.4.2(b), and 5.4.2(c) for chromate conversions of 50, 75, and 100%, respectively. The chloride conversion coatings were produced using Neostar™ black chloride. The general appearance of the chloride converted zinc is different from the chromate converted zinc. The chloride conversion gives rise to a dull surface consisting of smutty particles which can be rubbed off. The reflectance curve of chloride converted zinc coating is shown in Fig. 5.4.3.

The optimum values of solar absorptance and thermal emittance for chromate converted zinc and chloride converted zinc were 0.79 and 0.07, and 0.93 and 0.08, respectively. The chromate conversion produces a hard adherent coating while chloride conversion produces soft coating.

The process originated by Encheva (256) and followed by Vander Leij (251)

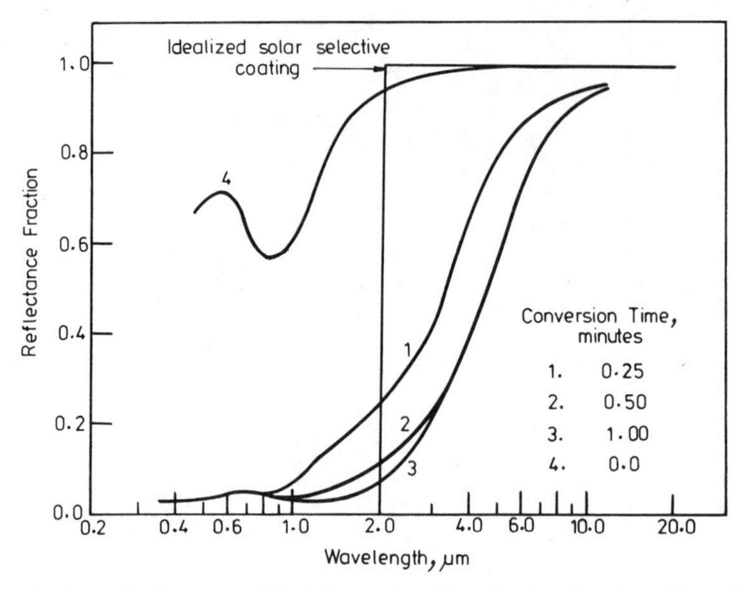

Fig. 5.4.3 Spectral reflectance of black zinc produced by chloride conversion. Adapted from McDonald and Curtis (1975) (Ref. 255).

for producing zinc oxide selective coating is described as follows: The clean steel samples were electroplated with bright zinc for about 30 min in an electrolyte containing 75 g/liter zinc cyanide, 125 g/liter sodium cyanide, 85 g/liter sodium hydroxide, 2 g/liter TNO organic brightner at pH > 13, at a temperature of 250°C with a current density of 4 A/dm^2. A zinc anode was used. For zinc oxide coating the samples were treated anodically with A.C. in a solution comprising of 25 g/liter NaOH, 20 g/liter $NaNO_3$ at a bath temperature of 25–40°C. The treatment time was 8–15 min and coating thickness was about 4 to 6 μm. Figure 5.4.4 shows the spectral reflectivity of black zinc oxide on zinc plated steel substrate prepared by anodic treatment in an electrolyte for different durations. The solar absorptance and thermal

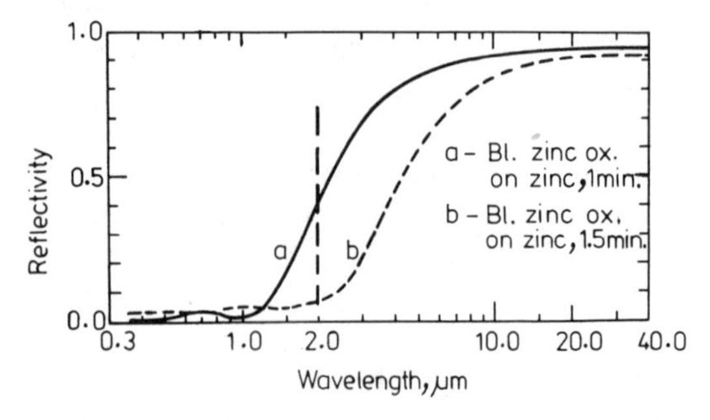

Fig. 5.4.4 Spectral reflectance of black zinc oxide on zinc plated steel prepared by anodic treatment in an electrolyte with 25 g/liter NaOH, 20 g/liter $NaNO_3$ at 40° C with 20 mA/cm^2. Adapted from Vander Leij (1978) (Ref. 252).

emittance values were of the order of 0.95 and 0.08, respectively. Vander Leij (252) reported that exposure of black zinc coating at low temperatures ($< 80°C$) and UV radiations does not show appreciable change in the optical properties, but exposure at high temperature increases the emittance. It is noticed that these coatings are quite corrosion resistant at low temperatures and therefore can be used in low-temperature solar collector applications.

5.4.3 *Colored Stainless-Steel Selective Surfaces*

Karlsson and Ribbing (258, 259), Smith (260), and Granziera (261) investigated selective properties of colored stainless-steel selective surface. In past the coloring process was intended for decorative purposes. The coloring of stainless steel was carried out by immersion in a hot solution containing appropriate concentrations of chromic and sulphuric acid (262) thus forming

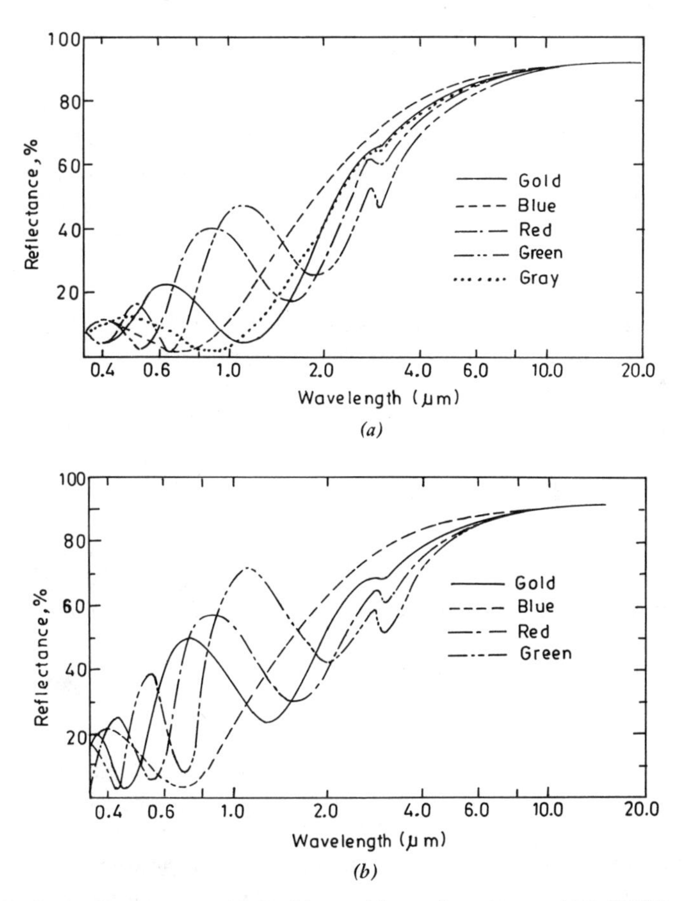

Fig. 5.4.5 Spectral reflectances (a) for blue, golden, red, and green SEL-STEEL and gray INCO-steel (b) blue, golden, red, and green INCO steel selective surfaces produced by chemical conversion. Adapted from Karlsson and Ribbing (1978) (Ref. 258).

Table 5.4.1 Solar Selective Properties of Chemically Converted Stainless-Steel Surfaces

Sample	α	$\varepsilon_{100°c}$	α/ε
Blue stainless steel (SEL)	0.90	0.1	9.0
Golden stainless steel (SEL)	0.86	0.1	8.6
Red stainless steel (SEL)	0.77	0.1	7.7
Green stainless steel (SEL)	0.78	0.1	7.8
Blue stainless steel (INCO)	0.83	0.1	8.3
Gray stainless steel (INCO)	0.91	0.1	9.1
Golden stainless steel (INCO)	0.68	0.1	6.8
Red stainless steel (INCO)	0.62	0.1	6.2
Green stainless steel (INCO)	0.63	0.1	6.3
Gray + Cr_2O_3	0.93	0.1	9.3
Gray + Si_3N_4	0.87	0.25	3.6

Source: Adapted from Karlsson and Ribbing (1978) (Ref. 258).

a film. However this film is soft and has limited applications because of softness. Evans et al. (263, 264) overcame this disadvantage by hardening process which was carried out by cathodic treatment in a similar chromic/sulphuric acid bath (262). The colored surface showed good corrosion and wear resistance and can be formed without any loss of adhesion or color intensity. Figure 5.4.5 (a, b) shows the room–temperature spectral reflectances of nine samples obtained from (258) stainless-steel equipment [SEL, London and international Nickel Company. Inc. (INCO) New York. The spectral reflectances beyond 7 μm are almost same in all the samples. Table 5.4.1 presents the solar selective properties of various colored stainless-steel selective surfaces. From Fig. 5.4.5 and Table 5.4.1 it is clear that blue steel shows the highest selectivity. Karlsson and Ribbing also studied the effect of post deposition heat treatment in vacuum and air.

Carver et al. (265) have proposed and successfully used a method to passivate molybdenum against oxidation by covering the surface with film of either Cr_2O_3, SiO_2, Si_2O_3, or Si_3N_4. The overcoated samples showed high absorptance for solar radiations.

5.4.4. *Alcoa 655 Selective Surface*

Powers et al. (266, 267) developed a new selective surface on aluminium by chemical conversion, which is called the Alcoa 655 process. Some of its advantages are good thermal stability, low cost, and availability on full-size solar collectors. After a conventional aluminium pretreatment cycle of cleaning, deoxidizing, etching, and desmutting, the coating is applied by immersing the plate in a dilute alkaline (pH 8 to 10) solution containing borate and silicate salts. Figure 5.4.6 is a flow chart for the preparation of

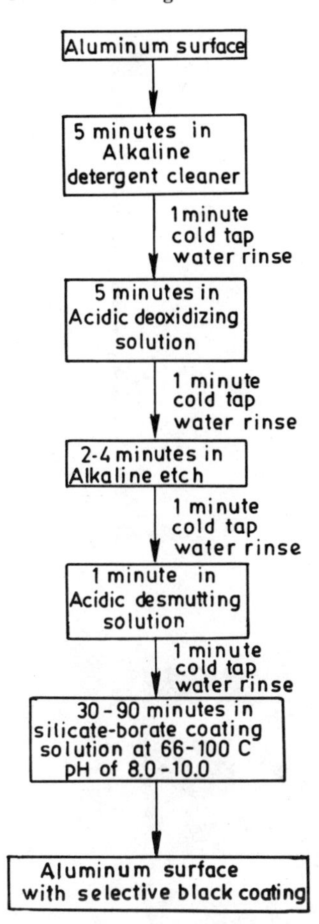

Fig. 5.4.6 Flow chart for the preparation of Alcoa 655 selective surfaces. Adapted from Powers et al. (1976) (Ref. 266).

the coatings. Typical immersion time and temperature are 45 min and 93°C, respectively. This chemically converted coating has a dark brown to black color. Electron probe microanalysis reveals the coating to be composed almost entirely of aluminium and oxygen. The thin (0.3–0.5 μm) film exhibits high solar absorptivity (0.93) and relatively low emissivity (0.35). Selectivity of the surface is attributed to metallic aluminium particles included in thin oxide film and to microcavities (etch pits) in the metal surface. Clusters of metallic aluminium atoms at anion vacancies in the film behave as color active sites and produce a dark black coloration. Etch pits associated with insoluble alloy constituents behave as microcavities with diameter in the 1 to 4 μm range. The metallic aluminium particles in the film are subject to oxidation and the reaction appears to be diffusion controlled. Figure 5.4.7 shows scanning electron micrographs of the surfaces of conversion coating and of the metal substrate after removal of the coating (267). Altenpohl (268) discussed tap water blackening of aluminium alloys, caused when these alloys contact certain heated, hard tap waters for many hours. He attributed

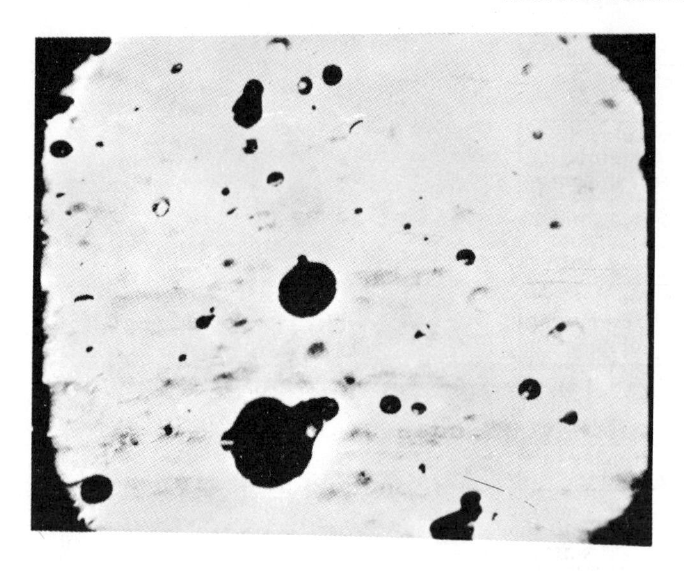

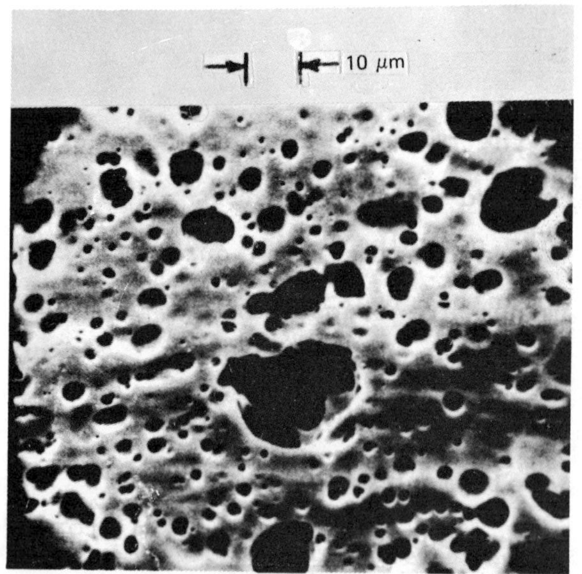

Fig. 5.4.7 Scanning electron micrographs of the surface of conversion coating (upper photograph) and the metal substrate (lower photograph) after removal of the coating. Adapted from Powers et al. (1976) (Ref. 266).

the blackening effect to a roughening of the metal surface by the attack of the hard water. Godard et al. (269) also described black staining of aluminium in hard waters. When hard water is boiled in an aluminium receptable, there is a rapid growth of surface oxide film during which particles of unoxidized aluminium are trapped in the film and cause it to acquire the dark color. The development of Alcoa 655 black coating is analogous to the above described methods. Analysis of Alcoa 655 films showed the presence of metallic aluminium mixed with the aluminium oxide in the region between the surface and oxide metal interface. The surface analysis showed the presence in the film of the elements Al, O, Si, and Na. Silicon originates from the silicate in the film despite its presence in the solution as sodium tetraborate. The film formation scheme for the Alcoa 655 process is shown in Fig. 5.4.8. It has been observed that the process does not blacken high purity aluminium (267). Aluminium containing alloying constituents, e.g., insoluble, iron-bearing constituents, is necessary for effective blackening. These constituents at the metal surface behave as local galvanic cells when contacted with an electrolyte and provide electric fields within the oxide films. The following reactions are postulated for the Alcoa 655 process (267). Sodium metasilicate dissociates into orthosilicate ion, the most abundant species in aqueous silicate solutions, and sodium ion:

$$Na_2SiO_3 + 2OH^- \rightarrow SiO_4^{4-} + 2Na^+ + H_2O \tag{5.4.1}$$

Anodic and cathodic sites occur on the metal surface as a result of local

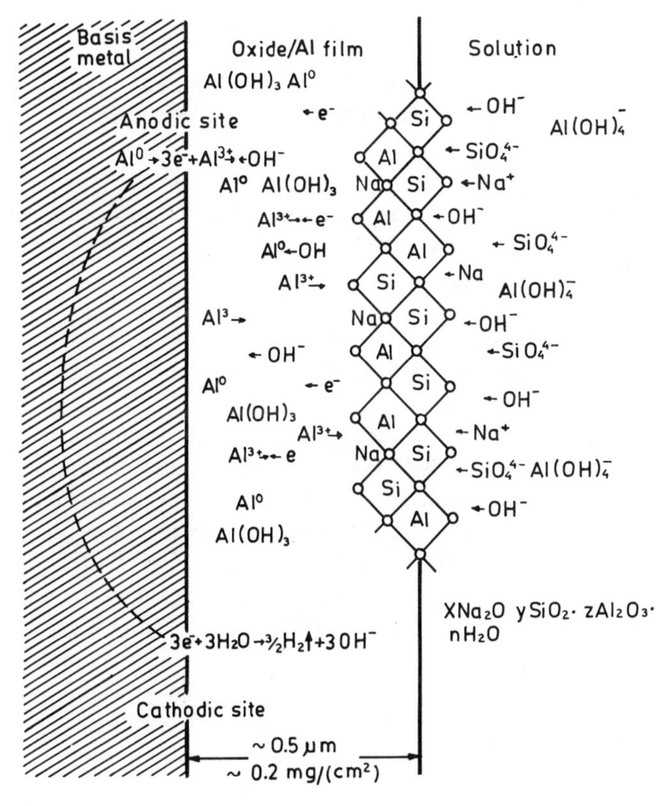

Fig. 5.4.8 Film formation scheme for Alcoa 655 selective surface. Adapted from Cochran and Powers (1976) (Ref. 267).

cell action. At anodic sites, aluminium ions are formed and tend to migrate out through the oxide film:

$$Al^0 \rightarrow Al^{3+} + 3e^- \quad (5.4.2)$$

At cathodic sites on the metal surface, water is reduced to hydrogen and hydroxyl ions:

$$3H_2O + 3e^- \rightarrow 1.5\,H_2\uparrow + 3\,OH^- \quad (5.4.3)$$

The overall reaction for dissolution of aluminium by the alkaline solution is formation of hydroxoaluminate ion and hydrogen:

$$Al + 6OH^- + 3H_2O \rightarrow Al(OH)_4^- + 1.5\,H_2\uparrow \quad (5.4.4)$$

When the solution becomes sufficiently supersaturated with respect to dissolved aluminium and has aged sufficiently, or is allowed to cool down, aluminium trihydroxide and boehmite precipitate, releasing hydroxyl ions:

$$Al(OH)_4^- \rightarrow Al(OH)_3\downarrow + OH^- \quad (5.4.5)$$

$$Al(OH)_4^- \rightarrow AlOOH\downarrow + OH^- + H_2O \quad (5.4.6)$$

Since dissolved silicate is a component of the solution, the actual precipitate obtained is probably a gelatinous zeolite of the general form:

$$X\,Na_2O\cdot Y SiO_2\cdot Z\,Al_2O_3\cdot n\,H_2O \quad (5.4.7)$$

Direct combination of aluminium ion and hydroxyl ion occurs within the film to form aluminium hydroxide:

$$Al^{3+} + 3OH^- \rightarrow Al(OH)_3\downarrow \quad (5.4.8)$$

Orthosilicate ion is discharged on the surface of the aluminium oxide film releasing its electrons from anodic sites within the film. Aluminium ions substitute for some of the silicon ions in the SiO_4 tetrahedra, producing a relatively insoluble aluminium silicate layer. Negative charges arising from the substitution of Al^{3+} for Si^{4+} result in cation vacancies and a negative space charge in the outer region of the film. Sodium ions from the solution enter the silicate layer to partially satisfy these negative charges. The aluminium layers tend to block penetration of the film by OH^- ions. This inhibiting effect of the silicate layer results in a deficiency of oxygen-bearing anions within the film. Consequently, anion vacancies and a positive space charge develop in the film near the metal/film interface. Aluminium ions migrating outward are starved for oxygen needed to form aluminium oxide or hydroxide. The excess aluminium ions share electrons with clusters of anion vancancies and are neutralized to metallic aluminium:

$$Al^{3+} + 3e \rightarrow Al^0 \quad (5.4.9)$$

The clusters of aluminium atoms and anion vacancies behave as color-

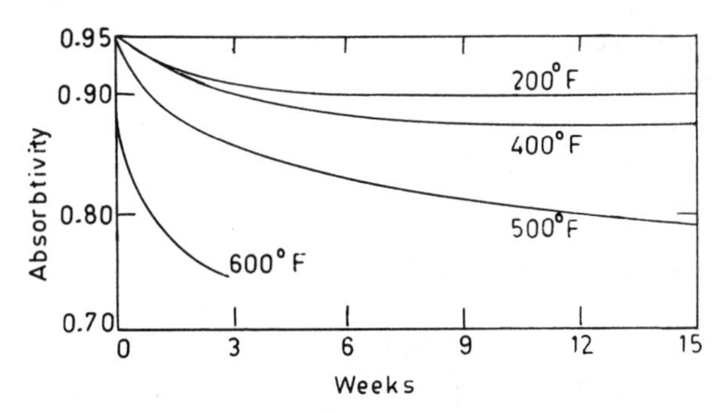

Fig. 5.4.9 Solar absorptance of Alcoa 655 selective surfaces heat treated for different durations. Adapted from Powers et al. (1976) (Ref. 266).

active sites, giving rise to the brown to black coloration of the film. The growth of the Alcoa 655 film is diffusion controlled and is self-limiting.

Powers et al. (266) have made measurements of solar absorptance and thermal emittance after exposure to elevated temperatures to determine thermal stability. Figure 5.4.9 shows solar absorptance for heat-treated samples for various durations. After exposure of 15 weeks at 200°C, the solar absorptance decreases from 0.95 to 0.87. To compare the effect of various covering materials on an Alcoa 655 coating, covered and uncovered specimens were exposed for 1000 hr (266). The covering materials were (1) window glass (3.2 and 6.4 mm thicknesses), (2) an 0.18 mm thick polyester film containing a UV absorber (Martin Processing Company), and (3) an

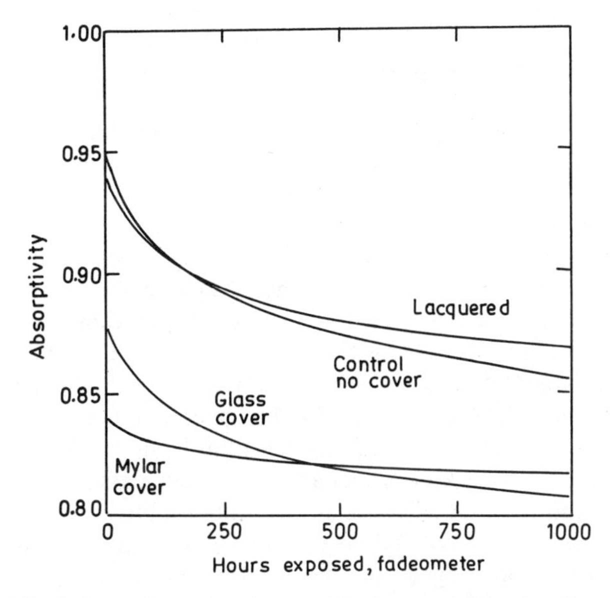

Fig. 5.4.10 Effect of covering materials on solar absorptance deterioration of Alcoa 655 selective surface. Adapted from Cochran and Powers (1976) (Ref. 267).

0.008 mm thick methacrylate lacquer film (Du Pont 1234). Figure 5.4.10 shows the effect of covering material on fading (loss in solar absorptance) of Alcoa 655 selective coatings. Further testing of Alcoa 655 is going on at the Alcoa Technical Center, USA.

5.5 Pure Semiconductors

The spectral selectivity can be obtained from an absorber–reflector tandem by overcoating an opaque metal having high thermal infrared reflectance with a thick film of semiconductor having an energy band gap from about 0.5 eV (2.5 μm) to 1.26 eV (1.0 μm) which would absorb solar radiations but is transparent to IR radiations. Such materials include Si (1.1 eV), Ge (0.7 eV), and PbS (0.4 eV). Semiconductor materials have high refractive indices n which give high reflectivities at air/semiconductor interface; e.g., PbS ($n = 4.1$) in vacuum has a normal reflectance of about 40%. The reflection coefficient can be reduced by proper thickness control to get destructive interference at solar maxima. The reflection coefficient can also be reduced by making a thin film of high porosity or by the application of antireflection coating. Kimoto and Nishida (270) described a gas evaporation technique for the porous deposits. In this technique the material is deposited in a gas atmosphere (0.1–1.0 torr) sufficient to cause vapor-phase nucleation of very fine particles (100–500 Å). In the next section we discuss the preparation techniques in brief and also the solar selective properties of Si, Ge, and PbS selective surfaces.

5.5.1 *Silicon and Germanium*

Mattox and Kominiak (271) prepared silicon and germanium films by an electron-beam evaporation system which was designed for electron-beam ion plating (272). This system allows the deposition chamber to be maintained at an argon pressure of up to 20 torr while the lower part of the electron-beam source can be maintained in the 10^{-4} torr range. Good adhesion of the silicon and germanium films to the metal surface was obtained by ion plating. These films were prepared by vacuum evaporation as well as by gas evaporation techniques. Deposition rates were about 1000 Å/min for vacuum evaporated silicon and germanium and about 300 Å/min for gas evaporated films. Figure 5.5.1 shows the spectral reflectance of vacuum evaporated and gas evaporated germanium films having different thicknesses deposited on 0.1 μm thick molybdenum film coated glass. The X-ray diffraction studies showed that silicon and germanium films were amorphous in nature. The gas evaporated and very thin vacuum evaporated films show a shift in the absorption edge to the shorter wavelengths compared to that of thick films. The solar absorptance increases from 0.56 to 0.61 in vacuum evaporated and 0.91 to 0.98 in gas evaporated germanium films having

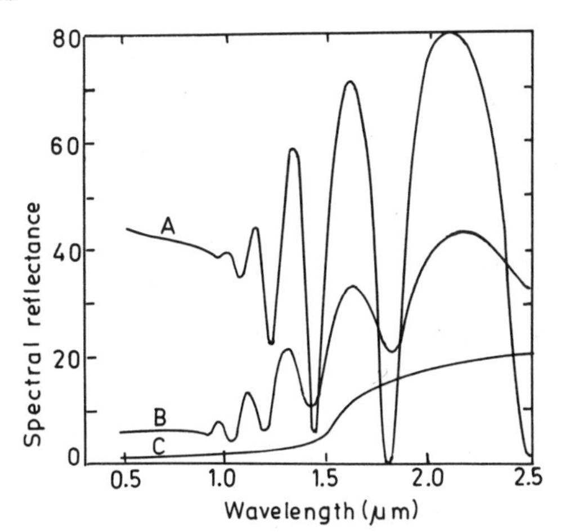

Fig. 5.5.1 Spectral reflectance of germanium film deposited on 1000 Å thick coated glass by vacuum evaporation and gas evaporation. Curve A: 5000 Å vacuum evaporated; curve B: 5000 A gas evaporated; curve C: 10,000 Å gas evaporated. Adapted from Mattox and Kominiak (1975) (Ref. 271).

0.5 μm and 1.0 μm thicknesses, respectively. Similar enhancement in solar absorptance was also observed in silicon films.

Recently Seraphin and coworkers have fabricated absorber–reflection tandem coatings using a CVD technique (273, 274). The complete system consists of three basic optical elements: the reflector, the absorber, and the antireflection layer. The three coatings were deposited on stainless-steel substrate which can withstand heat transfer at elevated temperatures. A barrier oxide was formed onto stainless-steel substrate by heating in air for 5 min at 500°C or in reactively evaporated chromium oxide or oxidized evaporated chromium to prevent diffusion between the substrate and the silver reflector film. The barrier oxide film can withstand the stresses induced by differential thermal expansion in cycling in the absorber stack up to 500°C. The evaporated silver layer was selected because of its low thermal emittance. Prasland et al. (275) have shown that silver films agglomerate at 300°C or above, but an overcoating of chromium oxide provides a complete stabilization against heating at temperatures above 800°C in a helium atmosphere. The chromium layer also prevents the diffusion between the silver layer and the silicon absorber layer. The silicon absorber layer was deposited by pyrolysis of silane in a helium atmosphere at 640°C. The antireflection coating was also deposited by the CVD method onto a silicon layer. A single layer of Si_3N_4 (700 Å thick) was deposited from a silane–ammonia mixture at 825°C. With the addition of an antireflection coating onto the silicon absorber layer, the solar absorptance increases from 0.40 to 0.75. The complete stack is shown in Fig. 5.5.2. Seraphin and Wells (276) described the method of fabrication of these stacks by CVD. The spectral reflectance of silicon tandem measured at 20 and 500°C is shown in Fig. 5.5.3.

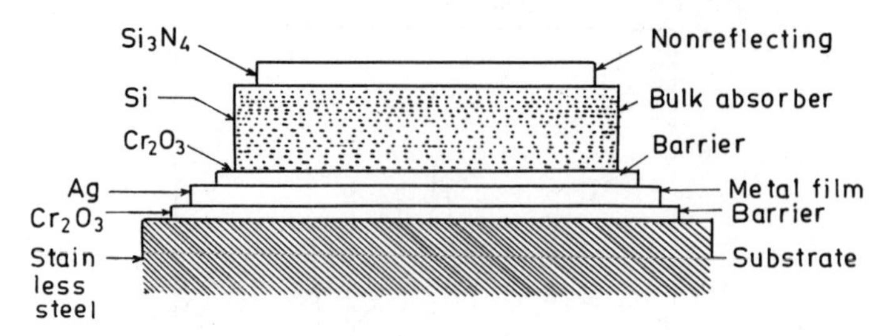

Fig. 5.5.2 Absorber–reflector tandem stack. Adapted from Seraphin (1974) (Ref. 83).

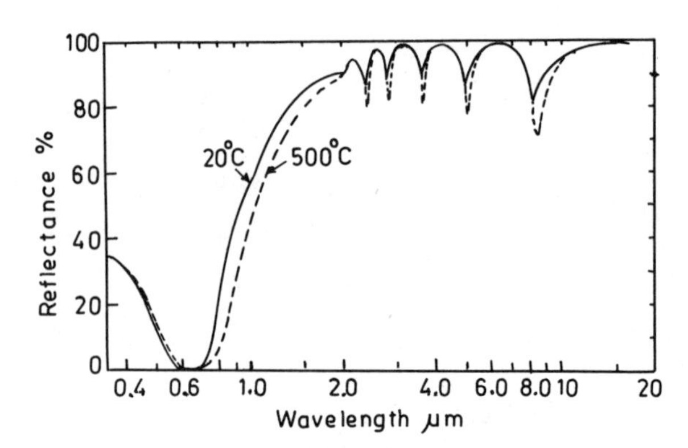

Fig. 5.5.3 The spectral reflectance of silicon tandem measured at 20 and 500 °C. Adapted from Seraphin (1975) (Ref. 276).

The spectral selectivity, i.e., the ratio of solar absorptance and thermal emittance, was 15 at room temperature and 12 at 500°C, a reduction of 10 to 12% over the room-temperature value. No degradation was observed after exposition in vacuum at 650°C for 20 hr or 200 times cycling at 500°C. No flaking or other mechanical damage occured during the life test experiments.

Recently Donnadieu and Seraphin (277) computed numerically the spectral reflectance as well as figure of merit of absorbing surfaces. The optimum values of parameters were selected and their dependence on thickness was studied. The silicon absorber layer converts solar photons into heat by absorbing them. Therefore with increasing thickness of the silicon layer the fraction of absorbed photon may be increased. However, if the silicon layer is too thick, it will contribute to the thermal emittance of the system in the IR. The reflectance of the system was calculated for values $d = 2.0$, 5.0, 10.0, and 15.0 μm of the thickness of the silicon absorber, with various types of antireflection layers. The results are listed in the Table 5.5.1 (adapted from Ref. 277) for various thicknesses and the three antireflection systems. It was found that with increasing thickness, the absorptance of the silicon layer increased. The greater path lengths lead to a shift of the absorption band edge to the long-wavelength side (Fig. 5.5.4). The emittance does not

Table 5.5.1 Results for the Various Absorber–Reflector Tandems Having Different Thicknesses
(Type I—Air, $n_1 = 1.43$, $n_2 = 1.62$, $n_3 = 1.81$, $n_4 = 2.00$; type II—Air, $n_1 = 1.43$, $n_2 = 1.82$, $n_3 = 2.2$, $n_4 = 2.79$)

Serial No.	Type	Thickness (μm)	Thickness Ge(μm)	Solar Absorp-tance	Emittance 300°K	600°K
1.	I	2.0	—	0.427	0.0418	0.0487
2.	II	2.0	—	0.467	0.0412	0.0482
3.	I	2.0	0.5	0.815	0.0385	0.0544
4.	II	2.0	0.5	0.890	0.0389	0.0545
5.	I	5.0	—	0.528	0.0384	0.0469
6.	II	5.0	—	0.572	0.0380	0.0465
7.	I	5.0	0.5	0.818	0.0410	0.0562
8.	II	5.0	0.5	0.892	0.0411	0.0562
9.	I	10.0	—	0.575	0.0376	0.0469
10.	II	10.0	—	0.630	0.0378	0.0469
11.	I	10.0	0.5	0.818	0.0434	0.0582
12.	II	10.0	0.5	0.894	0.0434	0.0581

Source: Adapted from Donnadieu and Seraphin (1978) (Ref. 277).

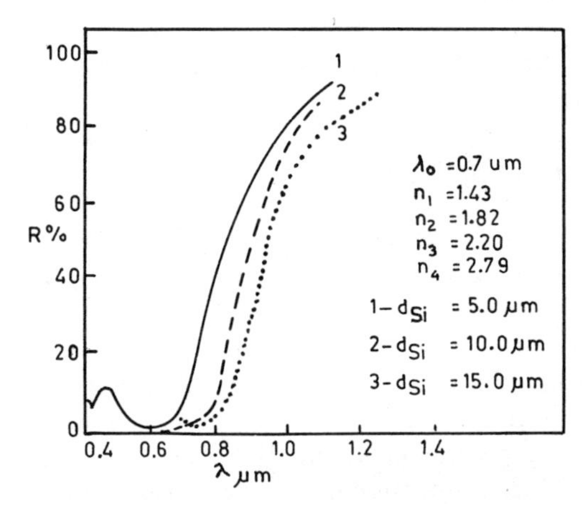

Fig. 5.5.4 Calculated spectral reflectance of silicon films having different thicknesses. Adapted from Donnadieu and Seraphin (1978) (Ref. 277).

increase with thickness up to 10 μm but it drops to smaller values on further increase in the thickness.

Donnadieu and Seraphin (277) suggested that some improvement can be made by adding a thin germanium layer to the silicon absorber. Germanium extends its intrinsic absorption into the IR which in turn increases the solar absorptance. The results of silicon layer having different thicknesses (2, 5, and

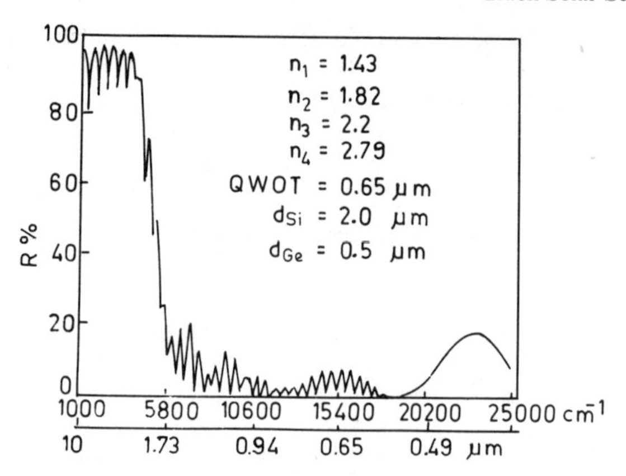

Fig. 5.5.5 Calculated spectral reflectance of an absorber–reflector stack in which 2.0 μm thick Si layer was overcoated with 0.5 μm Ge film. Adapted from Donnadieu and Seraphin (1978) (Ref. 277).

10 μm) overcoated with a 0.5 μm thick germanium layer are listed in the Table 5.5.1. Figure 5.5.5 shows spectral reflectance of 2 μm thick silicon layer overcoated with 0.5 μm germanium layer and antireflection layer. From Table 5.5.1 it is clear that an addition of a 0.5 μm thick germanium layer enhances the solar absorptance as well as the figure of merit. Best results were obtained in a tandem stack constituted of an absorber consisting of 0.5 μm germanium on a silver reflector overlaid by 2.0 μm silicon carrying a quadrupole antireflection sequence in which the refractive index varies from 2.79 to 1.43.

5.5.2 *Lead Sulfide*

Lead sulfide is another semiconductor, having a band gap of 0.4 eV (or 2.5 μm), which makes an absorber–reflector tandem of high selectivity (278, 279). High absorption in the solar region can be accomplished with a lead sulfide film that is thin compared to that of silicon and germanium. The reduction in the film thickness is due to high absorption coefficient because it is a direct band gap semiconductor where as silicon and germanium both are indirect band gap semiconductors. Williams et al. (278) in 1963 showed theoretically that high absorption in solar region and low thermal emissivity can be achieved in nonhomogeneous PbS coatings having high voids fraction ratio, i.e., a low refractive index. Figure 5.5.6 illustrates the effect of the void fraction on the refractive index of porous PbS, calculated from Garnett theory. In case of porous PbS, the refractive index can be lowered to 1.8 at 50% void fraction and to 1.2 at 80% void fraction provided the structural discontinuities are small compared to the radiation wavelength. McMahon and Jasperson (279) deposited PbS films on an Al substrate in a standard vacuum ($\approx 10^{-6}$ torr) by the evaporation technique. They reported solar absorptance of nearly unity and selectivity α_s/ε_T of 43 at room tempera-

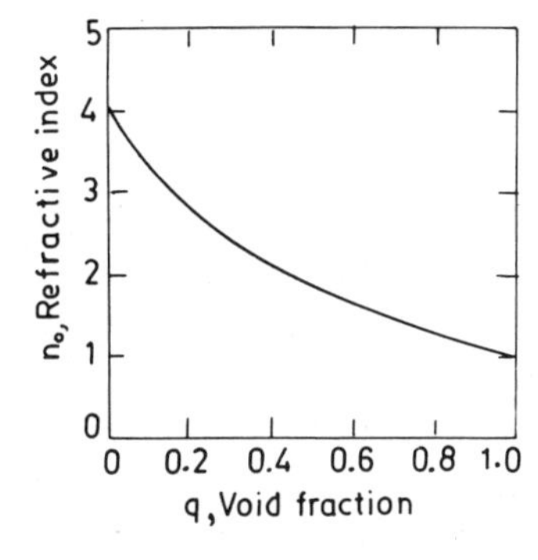

Fig. 5.5.6 Dependence of refractive index on void fraction ratio of porous PbS films calculated from Garnett's theory. Adapted from Williams et al. (1963) (Ref. 278).

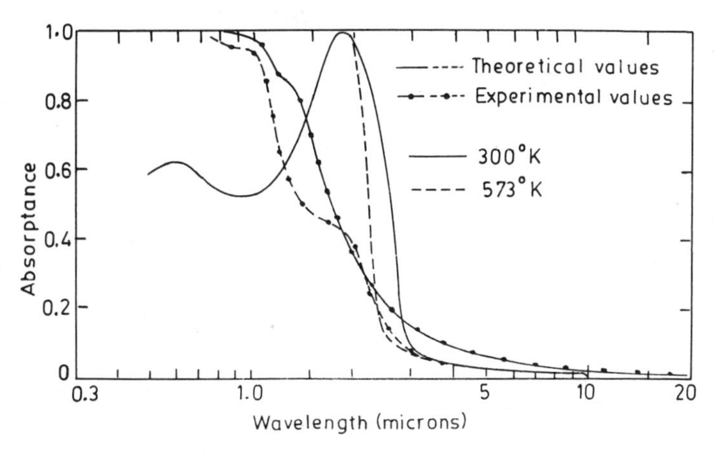

Fig. 5.5.7 Solar absorptance of PbS–Al system having 0.12 and 0.11 μm thick vacuum evaporated PbS film. Adapted from McMahon and Jasperson (1974) (Ref. 279).

ture without any aid of antireflection coating. Figure 5.5.7 shows the spectral absorptance of PbS–Al systems having 0.11 and 0.12 μm thick PbS films. A comparison with theoretically calculated 1000 Å thick PbS–Al system is also shown. The values of optical constants are taken from the standard literature.

Mattox and Kominiak (271) and Marchini and Gandy (280) studied the effect of PbS film thickness on the solar selective properties of PbS–metal system. Mattox and Kominak (271) produced PbS films by vacuum evaporation with a 30 Å/min rate. They observed that at higher deposition rates, the PbS film gave sooty deposits which were nonadherent. Figure 5.5.8 shows the effect of film thickness on the spectral reflectance of PbS films (271). The solar absorptance and thermal emittance both increases with thickness of PbS film. Figure 5.5.9 gives an idea of variation in selectivity with film thickness (280). Figure 5.5.9 shows both experimental and theoretically calculated values. A selectivity α_s/ε_T of 30 was noticed in 0.05 μm thick PbS film. Theoretical values of solar absorptance and thermal emittance were

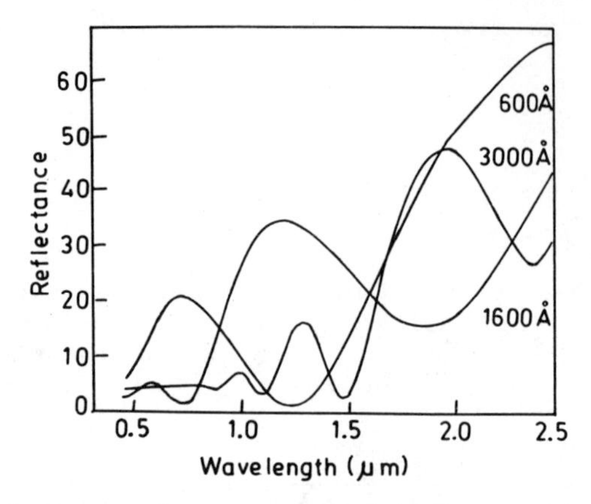

Fig. 5.5.8 Spectral reflectance of vacuum evaporated PbS film on molybdenum coated glass. Adapted from Mattox and Kominiak (1975) (Ref. 271).

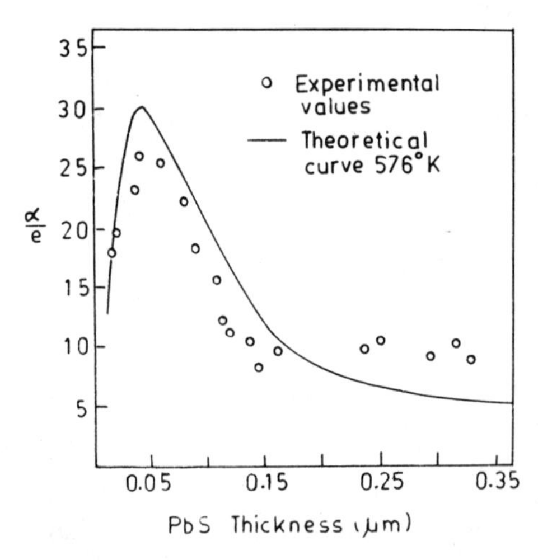

Fig. 5.5.9 Comparison of experiment and theoretical values of α/ε at 300°C as a function of thickness for PbS films. Adapted from Marchini and Gandy (1978) (Ref. 280).

calculated by using the selected ordinate method of Olson (281). The difference between experimental and theoretical values is attributed to the difference in surface structure and grain size of PbS films.

Recently Gupta et al. (282, 283) developed a low-cost technique for large-area selective surface. In this technique, the Al–PbS coatings were produced by a spray pyrolysis process. The solution sprayed PbS selective coatings were prepared on Al painted aluminium or galvanized-iron sheets by a suitable design of a double nozzle sprayer. The Al painted Al sheet or the galvanized-iron sheets were prepared by spraying Al paint onto the Al or galvanized-iron sheet. The PbS spray setup is described elsewhere (283).

The spray liquid was a 0.1 *M* solution of lead acetate and thiourea in water mixed in different mole ratios. The coatings prepared from solutions having different molar ratios are denoted by 1L1T, 1L2T, etc. where 1L1T stands for one part lead acetate and one part thiourea solutions of the same concentration. The solution was connected to the spray nozzle and sprayed onto substrates whose temperature was controlled by a uniformly heated hot plate. The liquid flow rate was monitored by a flowmeter and was controlled by air pressure. The spray mixture was blown onto hot substrate in a well-ventilated hood. Each spraying period was followed by a 2 min waiting period to prevent excessive cooling of hot substrate. After each spraying application, the substrate was rotated clockwise 90° to obtain a uniform thickness all over the surface. The important controlling parameters for the coating prepared by this method are (1) substrate temperature, (2) concentration and mole ratios of starting solutions, and (3) distance between spray nozzle and hot substrate. Coatings used in the present investigation were prepared in an air atmosphere, the spray rate was 5 ml/min, and the distance between spray nozzle and substrate was 45 cm. All the parameters were kept constant throughout the investigations. The thickness of the coating obtained was quite uniform and free of pinholes throughout each sample and the coatings grown to the same thickness have absorption spectra that vary among themselves by less than 0.01 in both high- and low-absorption regions. The PbS surface is stable at room temperature and has good adhesive properties. The solar absorptance and thermal emittance

Table 5.5.2 Solar Selective Properties Al–PbS Surfaces

Serial No.	Ref.	Material	Substrate	Preparation Technique	α_s	ε
1.	271	Al–PbS	Al Sheet	Evaporation	0.90–0.95	0.015–0.022
2.	280	PbS	Al Sheet	Evaporation	0.94	—
3.	282, 283	PbS	Al Sheet	1 L2T sprayed	0.80	0.19
4.	282, 283	PbS	Al Sheet	1 L3T sprayed	0.91	0.17
5.	282, 283	PbS	Al painted Al sheet	1 L3T sprayed	0.93	0.21
6.	282, 283	PbS	Al painted GI sheet	1 L3T sprayed	0.91	0.24
7.	282, 283	PbS	Al evaporated Al sheet	1 L3T sprayed	0.93	0.15
8.	282, 283	PbS	Al evaporated GI sheet	1 L3T sprayed	0.93	0.16
9.	282, 283	PbS	Al sheet	Vacuum evaporated	0.97	0.12
10.	282, 283	PbS	Al evaporated Al sheet	Vacuum evaporated	0.97	0.10
11.	282, 283	PbS	Al evaporated GI sheet	Vacuum evaporated	0.97	0.11

values have been listed in Table 5.5.2 for PbS–Al system prepared by various processes.

High absorptance and low emittance could only be obtained in the 1L3T specimen, i.e., the sample with one part lead acetate and three parts thiourea. Therefore, it appears that for the complete conversion of lead acetate into lead sulphide, more than the stoichiometric amount of thiourea will be needed due to the escape of unreacted H_2S. However, at a very large proportion of thiourea, the rate of formation of NH_2CN may be more than its rate of thermal ejection from the substrate and thereby results in the poor crystallinity of the film. The optimum composition of the solution will thus obviously be governed by the substrate temperature and also the rate of spray. The reflectance at the air–PbS interface was reduced by properly adjusting the thickness to get destructive interference at the wavelength of solar maximum (278). The use of antireflection coatings or porous film can reduce the reflectance further. Al painted Al sheets with PbS of 1L3T composition gives $\alpha_s = 0.93$ and $\varepsilon_T = 0.21$. Al evaporated Al sheet with PbS vacuum evaporated gives $\alpha_s = 0.99$ and $\varepsilon = 0.10$.

5.6 Metal Silicide and Carbide Solar Selective Surfaces

Harding et al. (284–288) studied in detail the solar selective properties of D.C. reactively sputtered metal carbide (285) and metal silicide (284) films on metals. Similarly Blickensderfor et al. (290) studied RF reactively sputtered metal carbide and metal nitride surfaces. The refractory nature and low vapor pressure of metal carbides (291) and metal silicides (289) enhances the stability at high temperatures of these selective surfaces.

The iron, stainless-steel, tungsten, chromium, etc., carbide and silicide

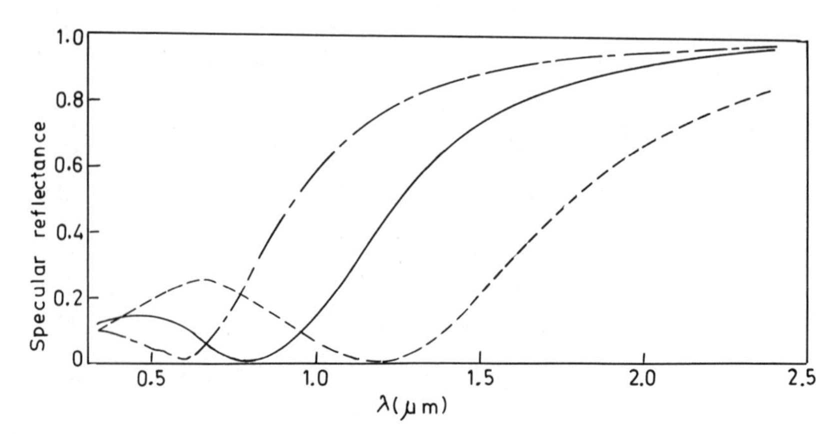

Fig. 5.6.1 Spectral reflectances for three thicknesses of iron carbide selective surfaces on bulk copper. Long-and short-dashed line, sputtering time = 5 min; solid line, sputtering time = 7 min; broken line, sputtering time = 10 min. Adapted from Geoffrey et al. (1978) (Ref. 288).

films were prepared by Harding et al. (284–287) by D.C. reactive sputtering of pure metals or metal mixtures in a gas mixture containing 2% methane or silane in argon. The films were sputtered under a pressure of 35 Pa, with cathode voltage 1 kV, current density 20 A/m^2 and a deposition rate of 0.04 $\mu m/min$ on bulk copper substrate. For optimum solar selective properties of homogeneous films, the deposition conditions were adjusted to produce a 0.09 μm thick film and electrical resistance of 5 to 100 kohm/sq. for metal carbide and 2 to 20 kohms/sq. for metal silicide films. Films having low or high resistance exhibit high solar reflectance which decreases the solar absorptance value. Figure 5.6.1 shows the spectral reflectance of iron carbide selective surface deposited onto bulk copper. The films, which have been prepared onto 0.2 μm thick sputtered copper deposited on glass slide, showed slightly high values of solar absorptance because they have lower film reflectance than the bulk copper substrates. The spectral reflectance for three iron silicide films of resistance 10–20 kohm/sq. on bulk copper is shown in Fig. 5.6.2. The optimum solar absorptance of 0.78 and thermal emittance of 0.025 at room temperature were observed in 0.9 μm thick iron silicide sputtered film.

Iron carbide and iron silicide films deposited on bulk copper and sputtered copper have been treated in air at 150 and 250°C. Table 5.6.1 lists the solar absorptance and thermal emittance of metal silicide and carbide on bulk copper and sputtered copper substrates (284, 285). The films deposited on bulk copper and sputtered copper substrates have shown different aging effects. After 250 hr at 250°C no deterioration was observed for surfaces containing bulk copper, while surfaces containing sputtered copper deteriorated slightly.

Harding (285) reported that the solar absorptance can be enhanced by making a multilayer stack of pure metal films and metal carbide films. Absorptance up to 0.90 has been reported for metal carbide selective surface prepared by depositing 0.05 μm thick metal film and 0.06 μm thick metal

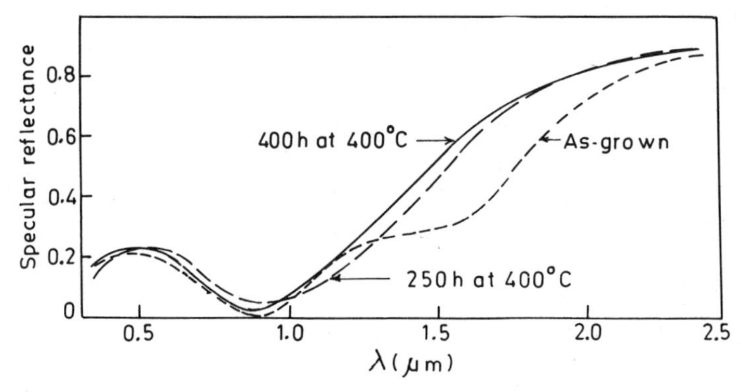

Fig. 5.6.2 Spectral reflectance of iron silicide films on bulk copper. Adapted from Harding (1978) (Ref. 284).

Table 5.6.1 Solar Absorptance and Thermal Emittance of Metal Silicide and Carbide films on Bulk Copper and Sputtered Copper Substrates

Serial No.	Coating	Substrate	As-Grown α	As-Grown ε	Heat Treatment Time in Hours	After Treating at 250°C α	After Treating at 250°C ε	After Treating at 400°C α	After Treating at 400°C ε
1	Chromium	Bulk	0.765	0.02	300	0.75	0.02	0.76	0.022
	silicide	copper	0.77	0.026	300				
2	Iron	Bulk	0.78	0.03	650	0.785	0.03	0.78	0.02
	silicide	copper	0.765	0.025	650				
3	Molybdenum	Bulk	0.78	0.02	300	0.795	0.022	0.80	0.024
	silicide	copper	0.79	0.024	300				
4	Stainless-steel	Bulk	0.815	0.022	330	0.82	0.022	0.80	0.022
	silicide	copper	0.815	0.020	330				
5	Tantalum	Bulk	0.800	0.022	330	0.81	0.022	0.80	0.022
	Silicide	copper	0.76	0.022	330				
6	Titanium	Bulk	0.81	0.020	330	0.81	0.022	0.805	0.022
	Silicide	copper	0.80	0.022	330				
7	Tungsten	Bulk	0.755	0.025	330	0.775	0.022	0.805	0.027
	silicide	copper	0.755	0.022	330				
8	Chromium	Sputtered	0.835	0.026	300	0.805	0.022	0.78	0.026
	silicide	copper	0.835	0.028	300				
9	Iron	Sputtered	0.83	0.035	650	0.815	0.030	0.780	0.03
	silicide	copper	0.84	0.035	650				

10	Molybdenum silicide	Sputtered copper	0.81 0.83	0.026 0.026	300 300	0.85	0.026	0.87	0.026
11	Tantalum silicide	Sputtered copper	0.85 0.82	0.030 0.035	650 650	0.835	0.025	0.835	0.026
12	Stainless-steel silicide	Sputtered copper	0.84 0.835	0.030 0.027	650 650	0.825	0.022	0.835	0.025
13	Titanium silicide	Sputtered copper	0.85 0.86	0.030 0.030	650 330	0.84	0.029	0.835	0.025
14	Tungsten silicide	Sputtered copper	0.815 0.815	0.030 0.032	330 330	0.80	0.025	0.81	0.027
15	Stainless-steel silicide	Evaporated aluminium	0.82 0.83	0.030 0.026	320 320	0.845	0.032	0.865	0.035
16	Stainless-steel silicide	Evaporated nickel	0.87 0.85	0.045 0.050	320 320	0.875	0.045	0.865	0.055
17	Stainless-steel silicide	Bulk stainless steel	0.845 0.87	0.1 0.1	320 320	0.865	0.105	0.875	0.1
18	Chromium carbide	Bulk copper	0.80	0.02	120	0.80	0.02	0.80	0.02
19	Iron carbide	Bulk copper	0.80	0.02	120	0.80	0.02	0.80	0.02
20	Molybdenum carbide	Bulk copper	0.79	0.02	120	0.79	0.02	0.79	0.02
21	Nickel carbide	Bulk copper	0.80	0.02	120	0.80	0.02	0.72	0.02
22	Tantalum carbide	Bulk copper	0.80	0.02	120	0.75	0.02	0.70	0.02

Table 5.6.1 (*Continued*)

Serial No.	Coating	Substrate	As-Grown		Heat Treatment Time in Hours	After Treating at 250°C		After Treating at 400°C	
			α	ε		α	ε	α	ε
23	Tungsten carbide	Bulk copper	0.76	0.02	120	0.73	0.02	0.70	0.02
24	Iron and chromium + nickel carbide	Sputtered copper	0.85	0.035	120	0.84	0.03	0.84	0.03
25	Chromium carbide	Bulk sputtered metal	0.85	0.05	120	0.89	0.05	0.75	0.05
26	Iron carbide	Bulk sputtered metal	0.87	0.035	120	0.91	0.035	0.87	0.045
27	Molybdenum carbide	Bulk sputtered metal	0.90	0.035	120	0.90	0.035	0.87	0.03
28	Nickel carbide	Bulk sputtered metal	0.80	0.06	120	0.80	0.06	0.90	0.06
29	Tantalum carbide	Bulk sputtered metal	0.87	0.035	120	0.80	0.03	—	—
30	Tungsten carbide	Bulk sputtered metal	0.84	0.035	120	0.88	0.045	0.88	0.045

Sources: Nos. 1–24 from Ref. 284; nos: 25–30 from Ref. 285.

carbide film onto each other. Sowell and Mattox (292) observed a variation in the chromium-to-oxygen ratio from a high value at the metal substrate–chromium oxide interface to a lower value at the chromium oxide surface. This variation contributes to the high solar absorption of electroplated black chromes. Fan and Spura (293) obtained high solar absorptance for sputtered chromium oxide on nickel substrate by depositing a chromium rich film followed by a chromium oxide Cr_2O_3 film. Multilayer iron, stainless steel and titanium silicide films were produced by Harding (284). A solar absorptance of 0.89–0.93 and thermal emittance of 0.03–0.04 at room temperature were observed. These surfaces are slightly less stable at elevated temperature due to interdiffusion between the layers.

5.7 Powdered Semiconductor–Reflector Combinations

This catagory is a subclass of absorber–reflector tandems described in previous section. In this type the particulate semiconductor coating is dispersed or deposited onto a highly reflecting substrate. The feasibility of this type of coatings was first demonstrated by Williams, Lappin and Duffie (278) in 1963. They prepared PbS coatings consisting of semiconductor particles suspended in an appropriate vehicle. These coatings have various advantages such as ease of application and fabrication, low cost, large-area availability, and environmental durability. Several authors have investigated solar selective properties of semiconductor (207, 227) inorganic metal oxides (211) and organic blacks (221), and metal dust pigmented (294, 295) selective paints. Extensive research was carried out at Honeywell Corporation USA by Mar et al. (211) on paint coatings. Several types of solar selective paints are discussed in the following sections.

5.7.1 *Semiconductor Pigmented Selective Paints*

Williams and coworkers (278) prepared lead sulfide in various crystalline forms and sizes by a precipitation technique and then mixed it with silicon resin. The coatings were deposited on pure aluminium substrate by settling or spraying techniques. Upon complete settlings of the PbS particles, the water was removed, and the wet coatings were then dried in a nitrogen atmosphere. The lead sulfide used for coating preparation was precipitated from a reaction mixture containing lead nitrate and thioactamide. The particle size was controlled by proper selection of acidity, reactant concentration, and reaction time, and large-sized cubic crystals were obtained at a lower reactant concentration of 0.02 *M*. After preparation, the particles were washed, dried, and dispersed in Dow Corning silicon 805 resin by grinding in a mortar. Xylene was added to bring the silicon to a viscosity at which it could be sprayed and the coating was applied by spraying onto a high purity aluminium disc with a small atomizer. The total spectral reflectance of

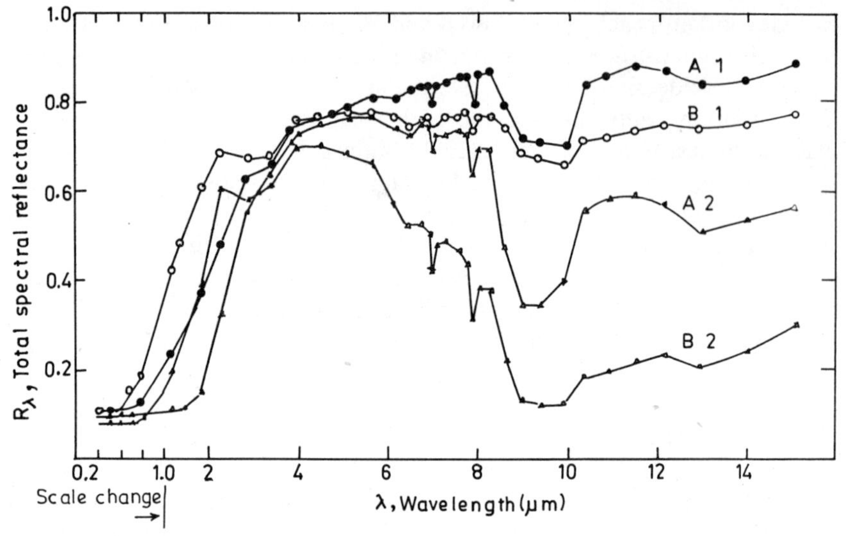

Fig. 5.7.1 Spectral reflectance of PbS–Si coatings deposited on pure aluminium sheet, at two particle sizes and two coating weights. A: 0.1 μm dendritic particles; B: 0.01 μm particles. (1): For a PbS coating weight of 0.17 mg/cm^2 and total weight of 0.22 mg/cm^2 (2): PbS coating weight of 0.56 mg/cm^2 and total weight of 0.72 mg/cm^2. Adapted from Williams et al. (1963) (Ref. 278).

PbS coatings with silicon resin is shown in Fig. 5.7.1 for 0.6 μm cubic and 0.1 μm dendritic coatings, both at a coating weight of 0.68 mg/cm^2 (278) Figure 5.7.2 illustrates the effects of the size of PbS crystals. These coatings included the silicon binder, with total coating weights of 0.22 mg/cm^2 and PbS weights of 0.17 mg/cm^2, and total coating weight of 0.72 mg/cm^2 and PbS weight of 0.56 mg/cm^2. The solar absorptance and

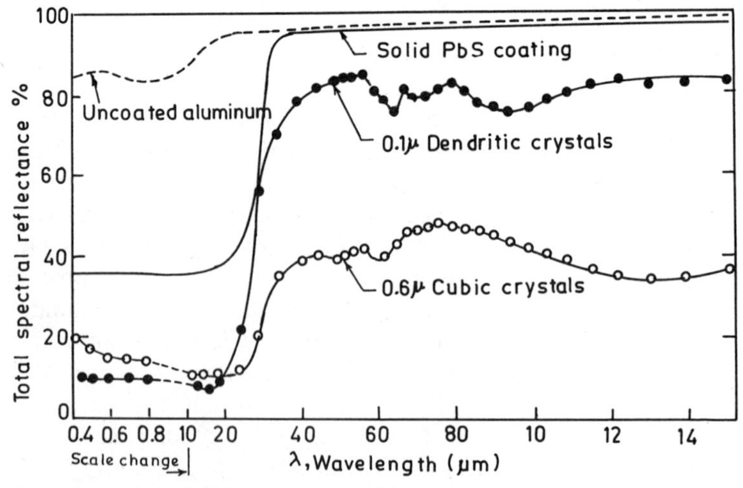

Fig. 5.7.2 Spectral reflectance of 0.68 mg/cm^2 of PbS on pure aluminium. The solid curve shows calculated from reflectance. Adapted from Williams et al. (1963) (Ref. 278).

thermal emittance increase from 0.84 to 0.92 and 0.19 to 0.71, respectively, as coating weight changes from 0.18 mg/cm^2 to 1.80 mg/cm^2. It appears that emittance is more strongly influenced by film thickness than the absorptance. Recently Pettit and Sowell (227) also prepared Si, Ge, and PbS semiconductor pigmented selective paints by mixing them in silicon resin. The PbS powder was prepared by the precipitation technique described above. The silicon and germanium particles were 325 mesh, available from commercial suppliers. The paints were prepared in a 3:1 volume ratio of particles

Table 5.7.1 Solar Selective Properties of Semiconductor Pigmented Selective Paints

Serial No.	Coating Thickness (mil)	Formulation	α_s	ε_T	Reference
1	—	32 PVC, PbS in DC 805	0.9	0.37	270
2	—	30 PVC PbS in DC 805	0.96	0.75	227
3	—	30 PVC Ge in DC 805	0.91	0.79	227
4	—	30 PVC Si in DC 805	0.83	0.70	227
5	0.80	30 PVC PbS in PP	0.92	0.80	211
6	0.21	30 PVC PbSe in EPDM	0.87	0.57	211
7	0.30	30 PVC PbS in EPDM	0.91	0.68	211
8	1.0	30 PVC CdTe in EPDM	0.88	0.80	211
9	0.20	30 PVC CdTe in EPDM	0.88	0.49	211
10	0.25	30 PVC Si in EPDM	0.79	0.56	211
11	0.21	30 PVC CdSe in EPDM	0.61	0.36	211
12	1.0	30 PVC Sb_2Se_3 in EPDM	0.80	0.53	211
13	0.25	30 PVC Bi_2S_3 in PP	0.85	0.61	211
14	0.90	30 PVC Bi_2S_3 in PP	0.84	0.69	211
15	0.35	30 PVC Bi_2S_3 in EPDM	0.85	0.67	211

to binder and were applied by spray onto a polished stainless-steel substrate. The paints were cured at 150°C for 2 hr. The solar absorptance were about 0.96, 0.91, and 0.83 in PbS, Ge, and Si selective paints, respectively, and total emittance was more than 0.70 in all the samples at room temperature.

Extensive studies on semiconductor pigmented selective paints have been made by Mar et al. (211). Approximately 30 semiconductors having high solar absorptance and IR transmittance with four binder materials in different pigment volume concentrations (PVC) were tried. The materials used as binders were olefin base polymers, polyethylene (PE), polypropylene (PP), ethylene–propylene material (EPM), ethylene–propylene–diene material (EPDM) and Dow Corning silicon resin. The olefin polymers are transparent in the IR but they are not thermally stable above 100°C. Silicon resins are less transparent in the IR than olefin polymers, but they are stable at high temperatures. Table 5.7.1 lists several semiconductor pigmented solar selective paints. The coatings having thickness more than 1 mil show a nonselective behavior. Therefore, in order to achieve selective solar optical properties, thicknesses should be much less than 1 mil.

5.7.2 *Inorganic Metal Oxides Pigmented Selective Paints*

Mar et al. and Lin and Zimmer (211, 221) investigated optical properties of various inorganic metal oxides like copper–chrome oxide, Cu–CrO_x Meteor 7890 (222), chrome–copper oxide, Cr–CuO_x (V–302 and F 2302) (296), iron manganese–copper oxide, FeMn–CuO_x (F-6331) (296), copper–chromium–manganese oxide $CuCrMnO_x$ (L-4128-12) (222), cobalt–iron–manganese oxide $CoFeMnO_x$ (L-3850-30) (222), copper–iron–manganese oxide, $CuFeMnO_x$ (L-3938-19) (222) and cobalt-iron oxides, $CoFeO_x$ (L-4128-45) (222), etc., mixed with different binder materials. The binders

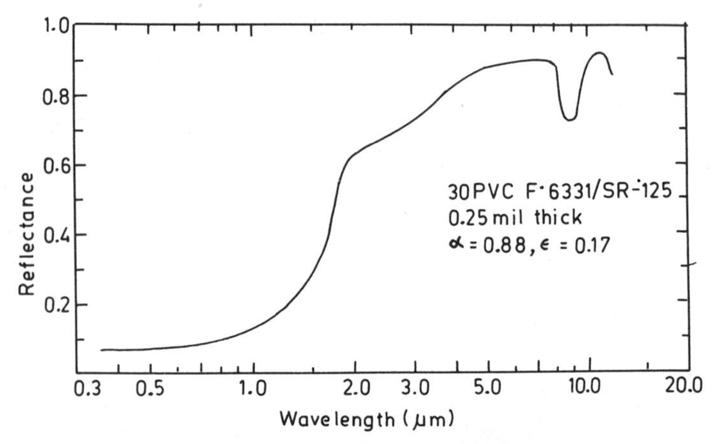

Fig. 5.7.3 Spectral reflectance of a thin inorganic selective paint coating. Adapted from Lin and Zimmer (1977) (Ref. 221).

Table 5.7.2 Solar Selective Properties of Inorganic-Oxide Pigmented Selective Paints

Serial No.	Coating Thickness (mil)	Formulation	α_s	ε_T
1	0.18	30 PVC Meteor 7890/EPDM	0.94	0.52
2	0.18	30 PVC V-302/EPDM	0.93	0.57
3	0.15	30 PVC F-6331/EPDM	0.96	0.56
4	0.12	30 PVC F-6331/SIL	0.94	0.40
5	1.0	30 PVC F-6331/PP	0.95	—
6	0.9	30 PVC F-6331/PE	0.94	—
7	0.06	30 PVC F-6331/PE	0.92	0.36
8	0.05	30 PVC F-6331/EPDM	0.90	0.24
9	0.30	30 PVC F-6331/SR-125[a]	0.93	0.44
10	0.30	30 PVC V-302/SR-125	0.85	0.44
11	0.30	30 PVC Meteor 7890/SR-125	0.89	0.47
12	0.24	30 PVC F-6331/SR-125	0.87	0.16
13	0.40	30 PVC L-4128-12/SR-125	0.92	0.68
14	0.40	30 PVC L-3850-30/SR-125	0.91	0.60
15	0.39	30 PVC L-3938-19/SR-125	0.92	0.70
16	0.40	30 PVC L-4128-45/SR-125	0.87	0.63

Source: Adapted from Mar et al. (1976) (Ref. 211).
[a] SR-125 is SR-125-Silicon Resin (Dow Corning).

used were PP, EPDM, and a silicon resin (SIL). The pigments were calcined metal oxide combinations. The paint coatings were applied onto polished and pure aluminium substrates and cured. Table 5.7.2 lists solar selective properties of several inorganic metal oxide pigmented selective paints. Paint coatings consisting of F-6331 pigment mixed with silicon resins (either DC 808 or SR-125), show promising values of solar absorptance ($\alpha_s = 0.92$) and thermal emittance ($\varepsilon_T = 0.13$). Figure 5.7.3 shows the spectral reflectance of 30 PVC F-6331/SR-125, 0.25 mil thick selective paints. The typical values of solar absorptance and thermal emittance are 0.88 and 0.17, respectively. From Table 5.7.2, it is clear that coatings using pigments F-6331 have better absorptance/emittance characteristics than coatings using pigments V-302 or meteor 7890 with the same PVC and coating thickness.

5.7.3 *Organic Black Pigmented Selective Paints*

Lin and Zimmer (221) produced black selective paints by spraying black organic pigments mixed with acrylic lacquer. Paint coating with 70 PVC of organic black were formulated. The black organic pigments are available with Ciba Geigy Corp. Ardsley N Y and Mobay Chemical Corp. Verona

Table 5.7.3 Solar Selective Properties of 70 PVC Organic Black Pigments and Acrylic Lacquer Binder Selective Paints.

Serial No.	Coating Thickness (mil)	Formulation [a]	α_s	ε_T
1	0.04	Orosol Black CN(CG)	0.62	0.12
2	0.10	Orosol Black CN(CG)	0.68	0.31
3	0.01	Orosol Black RL(CG)	0.57	0.05
4	0.04	Orosol Black RL(CG)	0.62	0.25
5	0.06	Orosol Black RL(CG)	0.64	0.30
6	0.23	Irgalite SN(CG)	—	0.68
7	0.16	Black 7984 (MCC)	—	0.56
8	0.11	Oil Black BN (MCC)	—	0.34
9	0.10	Lacquer Black (MCC)	—	0.46
10	0.14	Nigresine Base D(MCC)	—	0.44

Source: Adapted from Lin and Zimmer (1977) (Ref. 221).
[a](CG) = Ciba Geigy Corp.; (MCC) = Mobay Chemical Corp.

Dyituff Div. Union. NJ, both American commercial manufacturers. Various organic black pigments are listed in Table 5.7.3. From this table, it is clear that the paint coatings based on organic black pigments are capable of low emittance but with low solar absorptance. The transition from low reflectance to high reflectance takes place at 0.7 μm, which in turn results in a low solar absorptance value. Figure 5.7.4 shows a comparison in spectral reflectance between organic blacks and inorganic metal oxides pigmented selective paints. From Table 5.7.3 it is also clear that as the coating thickness increases,

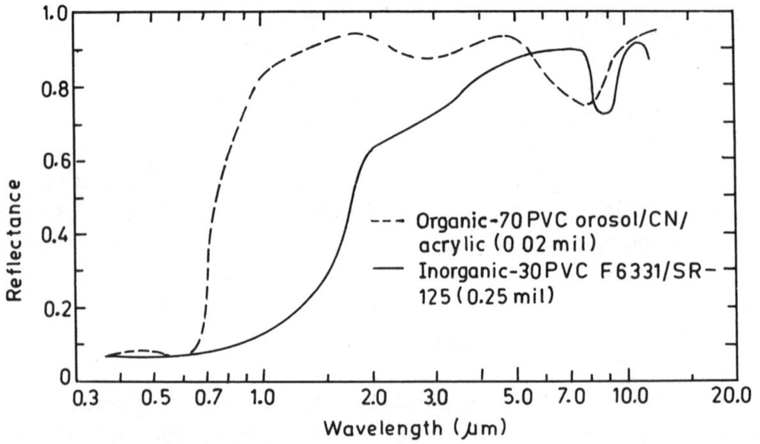

Fig. 5.7.4 Spectral reflectance of organic black and inorganic metaloxide pigmented selective paint coatings. Adapted from Lin and Zimmer (1977) (Ref. 221).

the emittance also increases. However, the increase in absorptance is very much less. The low solar absorptance may be enhanced by mixing same metal oxide in the organic black pigments. Table 5.7.3 lists the solar selective properties of organic black pigmented selective paints having different thicknesses and PVC ratio of 70.

5.7.4 *Metal Dust Pigmented Selective Paints*

Recently Telkes (294) followed a new approach to low-cost large-area selective surfaces by coating reflective metal particles with a layer of black selective materials like CuO, CuS + PbS, etc. The coated particles were mixed in a suitable binder material and applied onto the substrate. In the work of Gupta et al. (295), all the black coatings were deposited on zinc metal dust. The zinc powder was first cleaned to remove grease and any zinc oxide film, if present. A dilute solution of sodium carbonate was used to remove grease. A solution of 5% HCl was used to remove zinc oxide layer.

For CuO deposition, a solution was prepared by dissolving 10–30 g of $CuSO_4 \cdot 5H_2O$ in 100 ml of boiling distilled water and adding 0.05 *N* NaOH. Upon addition of NaOH, a precipitate at first forms and then dissolves. 100 g zinc powder were then added in a boiled solution. In a few seconds, the zinc metal powder became coated with an adherent black coating. The solution was decanted from the coated particles and then cleaned with distilled water several times using a sintered funnel. The coated particles were then dried in air. For the deposition of copper oxide on aluminium, two more types of solutions were prepared. One type consists of copper nitrate, concentrated nitric acid, and potassium permanganate in 25, 3, 15 g/liter concentrations in water. The deposition of aluminium metal powder (325 mesh) was carried out at 85–90°C for 15 min. In the other type, the bath comprises $NaClO_2$ and NaOH solutions in water. The deposition was carried out for 10 min at 100°C. After deposition of the copper oxide coating on aluminium metal powder, it was cleaned with distilled or deionized water and dried in air.

For CuS deposition, separate solutions of copper tartarate, tartaric acid, and NaOH were prepared in 15, 20, and 30 g/liter concentrations and mixed at room temperature. The cleaned zinc metal dust was added at room temperature. The mixture was rapidly stirred during the deposition. The coated particles were decanted from the solution and cleaned several times with distilled and deionized water, and then dried in air. The coated powder so obtained was treated with a dilute solution of sulphur in CS_2 to convert the coating into sulphide form for CuS + PbS deposition. The separate solutions of sodium thiosulfate, lead acetate, potassium hydrogen tartarate, and $CuSO_4 \cdot 5H_2O$ were prepared in 240, 25, 30, and 20 g/liter concentrations, respectively, and were mixed together. The deposition was carried out by mixing 50 g zinc dust with 100 ml of solution at 50°C. An adherent coating was deposited in seconds time onto zinc metal dust.

Table 5.7.4 Solar Selective Properties of Metal Dust Pigmented Selective Paint

Serial No.	Materials (Pigment + Silicon Binder)	Substrate	Preparation Technique	α_s	ε_T
1	Cupric oxide coating on Zn powder by electrolessdeposition[a]	Aluminium sheet	By paint brush	0.95	0.42
2	CuS coating on Zn powder by electrolessdeposition[a]	Aluminium sheet	By paint brush	0.94	0.51
3	CuS + PbS coating on Zn powder by electrolessdeposition[a]	Aluminium sheet	By paint brush	0.96	0.49
4	CuO coated Zn powder[b]	Aluminium sheet	Spraying	0.94	—

[a] *Source:* Ref. 295.
[b] *Source:* Ref. 294.

The coated metal dust was mixed in a resin and applied onto Al, GI, and copper sheet by spraying or painting by brush. The coatings have excellent resistance to extremes of temperature, moisture, and weathering. They can withstand 16 hr exposure at 250°C to 50% hydrochloric, nitric, and sulfuric acids and to 20% sodium hydrogen solutions. The pigment to binder material ratio was 3:1. The curing was carried out at room temperature for 12 hr. The selective paint was applied to a clean, dry surface at room temperature. The substrate should be clean to prevent poor adhesion before the paint application.

Telkes (294) found solar absorptance to be 0.94 but no emittance data was reported. The paint coatings prepared by Telkes cured at 120°C while the coatings prepared by Gupta et al. (295) cured at room temperature. The result of solar absorptivity and thermal emissivity reported by Gupta et al. (295) are listed in Table 5.7.4. Cupric oxide coated zinc dust pigmented solar selective paint gives an α_s of 0.95 and an ε_T of 0.42. The cupric sulfide and cupric sulfide + lead sulfide coatings give α_s as 0.94 and ε_T as 0.51. The higher emissivity is attributed to uncontrollable amount of black coating deposited onto fine metal particles.

5.8 Multilayer Interference Stacks

In absorber–reflector tandems the selective effect is caused by a single pass through the optically active medium, or the return pass after reflection by the underlying mirror surface, respectively. However, in case of multilayer

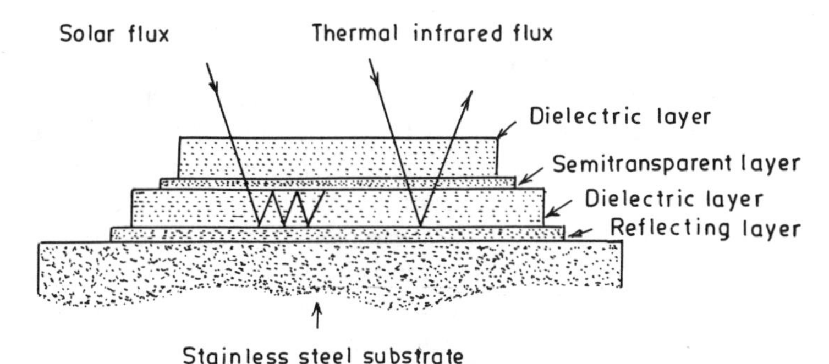

Fig. 5.8.1 A four-layer interference stack consisting of two quarter wave thick dielectric layers and thin semitransparent metal films. Adapted from Seraphin (1974) (Ref. 83).

interference stack the selective effect is a result of multiplicity of passes through the dielectric layer of the stack sandwiched between semitransparent and totally reflecting surface. Seraphin (83) prepared such an interference stack (Fig. 5.8.1). It depicts a four-layer interference stack comprised of two quarter-wave dielectric layers separated by thin semitransparent film. In this case it is not necessary for dielectric layer to have intrinsic absorption for the stack to be an effective absorber. The general characteristics of the four-layer interference stack are shown in Fig. 5.8.2. The first curve corresponds to the reflectivity of metal layer which has high reflectance in the IR region and slightly less reflectance in the visible region. By the addition of the second layer of dielectric material, the reflectance in the visible is reduced and the shape and the position of the curve is dependent on the dielectric layer thickness. The further addition of a third semitransparent metal layer reduces the reflectance in the visible region as shown by

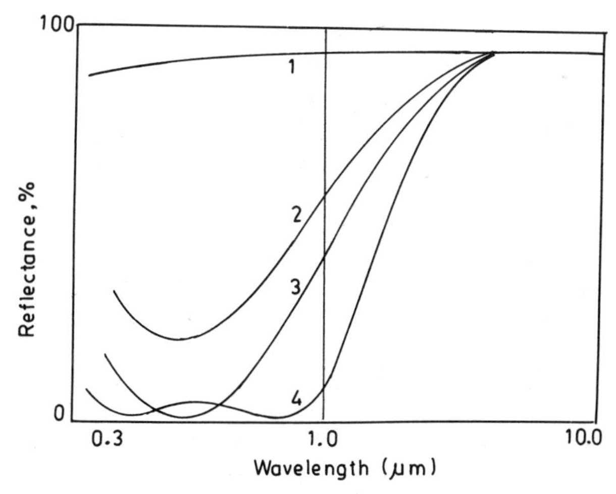

Fig. 5.8.2 Spectral reflectance for a four-layer interference stack as successive layers are added. Adapted from Seraphin (1974) (Ref. 83).

curve 3. The final fourth dielectric layer increases the absorption in the visible region and broadens the region of high absorption. The basic physics of multilayer stack is discussed by Meinel and Meinel (206). The equations for the film thickness necessary to achieve the maximum and minimum values of reflectivity and the conditions of zero reflectance for stacks with one or two dielectric films over a metallic substrate are derived by Park (297) and Schmidt et al. (298, 299). Park (297) also derived equations necessary to achieve the required reflection characteristics for multilayer stacks. Sharafi and Mukminova (300, 301) also describe a procedure for computing the reflectance, absorptance, and transmittance for radiant energy in multilayer system with varying optical properties (i.e., differing reflection, absorption, and refraction coefficients). The multilayer interference stacks are stable at elevated temperatures and exhibit high solar absorptance and low thermal emittance. Several interference stacks comprising different metals and dielectric layers such as Al_2O_3, SiO_2, CeO_2, ZnS, etc. have been discussed in the literature (299, 300).

Hottel and Unger (235) and Watson-Munro and Horwitz (302) produced copper oxide selective surface by deposition of droplets of black copper oxide over anodized aluminium and by treatment of a bright copper sheet with a 2:1 solution of NaOH and $NaClO_2$ for 3–10 min. Figure 5.8.3 shows the spectral specular reflectance of both types of selective surfaces. For comparison, the spectral reflectance of bulk copper is also shown.

Figure 5.8.4 shows the spectral reflectance of electrodeposited black nickel interference stacks reported by Tabor (232), Mar et al. (304), and Peterson and Ramsey (307). The coating by Tabor exhibits a double minimum for the interference film, around wavelengths of 0.4 and 1.0 μm. This coating consists of two layers of ZnS and NiS brought about by a simple change in plating current density during the electrodeposition process. Mar et al.

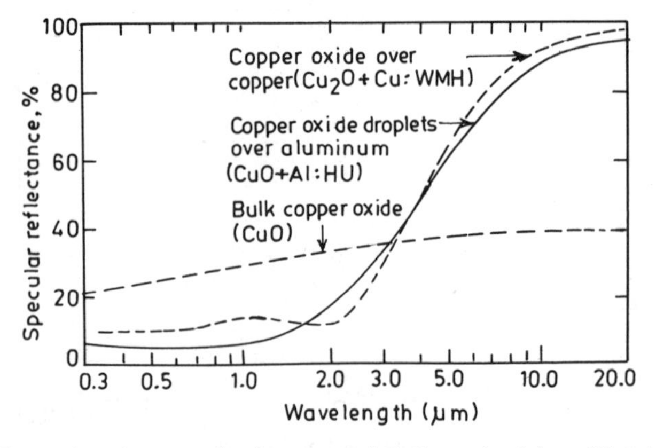

Fig. 5.8.3 Comparison in spectral reflectances of CuO on aluminium (Hottel and Unger) and Cu_2O on copper (Watson–Munro and Horwitz, 1975) and bulk CuO. Adapted from Meinel and Meinel (1977) (Ref. 206).

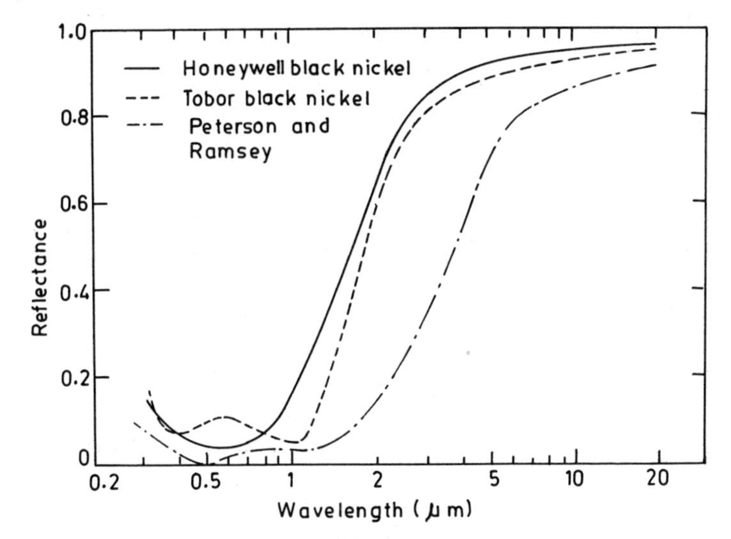

Fig. 5.8.4 Spectral reflectance of electrodeposited black nickel interference stacks reported by Tabor (232), Mar et al. (304), and Peterson and Ramsey (307).

(304) at Honeywell have shown an improvement in the solar absorptance value. The coating behavior also changes on post-plating heat treatment. The transition from absorptance to reflection moves towards longer wavelength region after treatment at 500°C for 15 hr. Therefore the transition of high absorptance to high reflection can be controlled by heat treatment after electrodeposition.

The Helio stacks investigated by Meinel et al. (305) at Helio Associates exhibit a fast rise in the reflectance, but the transition from high solar absorptance to high IR reflectance occurs in visible region. Therefore the net solar absorption is low, on the order of 0.70 to 0.75. The spectral reflectances is

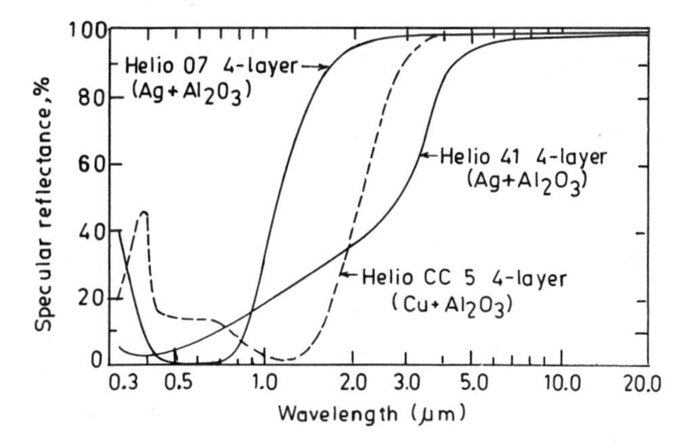

Fig. 5.8.5 Spectral reflectance of interference stacks by McKenney and Baauchamp using silver and copper based reflective films. Adapted from Meinel et al. (1974) (Ref. 305).

shown in Fig. 5.8.5 for various stacks. The transition can be shifted significantly by changing thickness and deposition conditions. A better response was obtained from the Al–SiO_2 and Helio CC-5 coating which has a steep transition with reasonably good solar absorption.

Schmidt and Park (303) at Honeywell Inc. have made extensive studies on metal–dielectric multilayer coatings, e.g., MgF_2–Mo–CeO_2, MgF_2–Mo–MgF_2–Mo–MgF_2, MgF_2–CeO_2–Mo–MgF_2–CeO_2, and Al_2O_3–Mo–Al_2O_3–Mo–Al_2O_3–Mo–Al_2O_3, etc. deposited on highly reflecting molybdenum substrate. The films were prepared by a vacuum evaporation method and the thickness was controlled by a quartz crystal thickness monitor. Figures 5.8.6 to 5.8.9 show spectral reflectance curves for MgF_2–Mo–CeO_2, MgF_2–Mo–MgF_2–Mo–MgF_2, MgF_2–CeO_2–Mo–MgF_2–CeO_2, and AMAMAMA on molybdenum multilayer coatings. Table 5.8.1 lists the solar absorptance and thermal emittance at two different temperatures for these coatings.

Recently Li et al. (306) reported solar selective properties of SiO_2–Mo–SiO_2, CeO_2–Mo–CeO_2 multilayer coatings prepared by a vacuum thermal evaporation method. Figures 5.8.10 and 5.8.11 show the spectral reflectance curves of SiO_2–Mo–SiO_2 and CeO_2–Mo–CeO_2 coatings applied on molybdenum foil. With the aid of two-layer antireflection coatings, the reflection in solar region can be reduced or the solar absorptance enhanced. Figure 5.8.10 shows that the transition edge from high solar

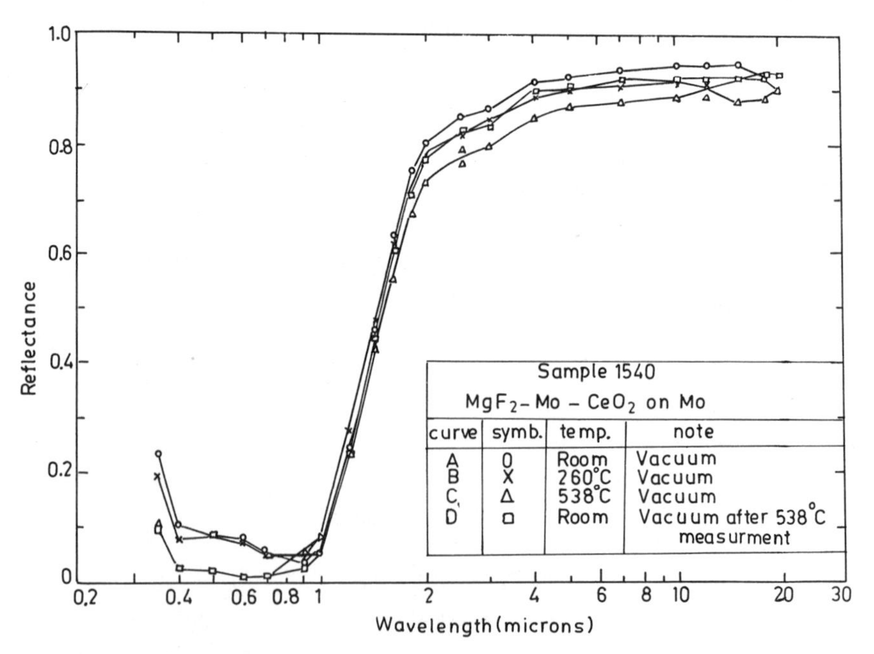

Fig. 5.8.6 Spectral reflectance of MgF_2–Mo–CeO_2 on Mo samples. Adapted from Schmidt and Park (1965) (Ref. 303).

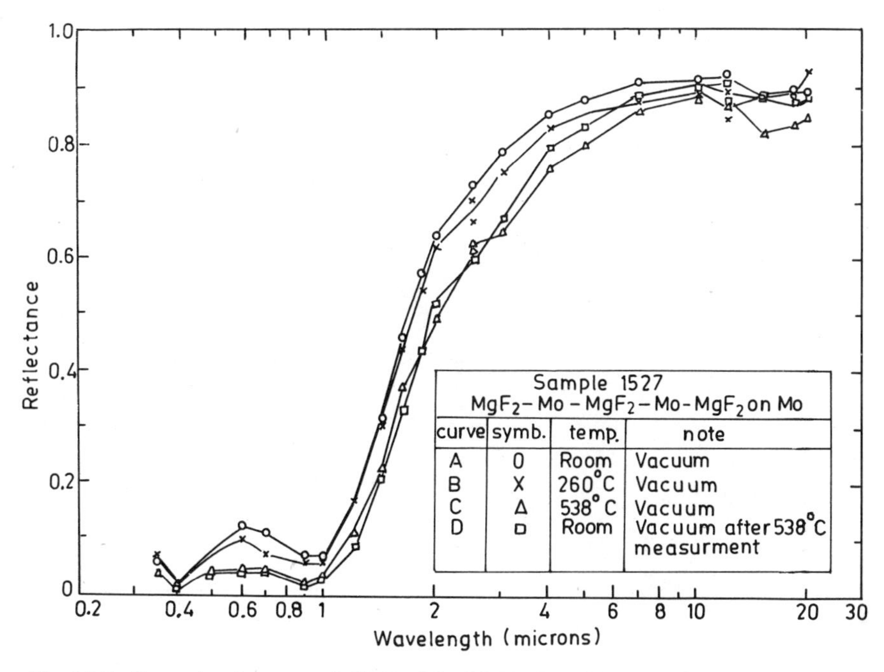

Fig. 5.8.7 Spectral reflectance of MgF_2–Mo–MgF_2–Mo–MgF_2 on Mo. Adapted from Schmidt and Park (1965) (Ref. 303).

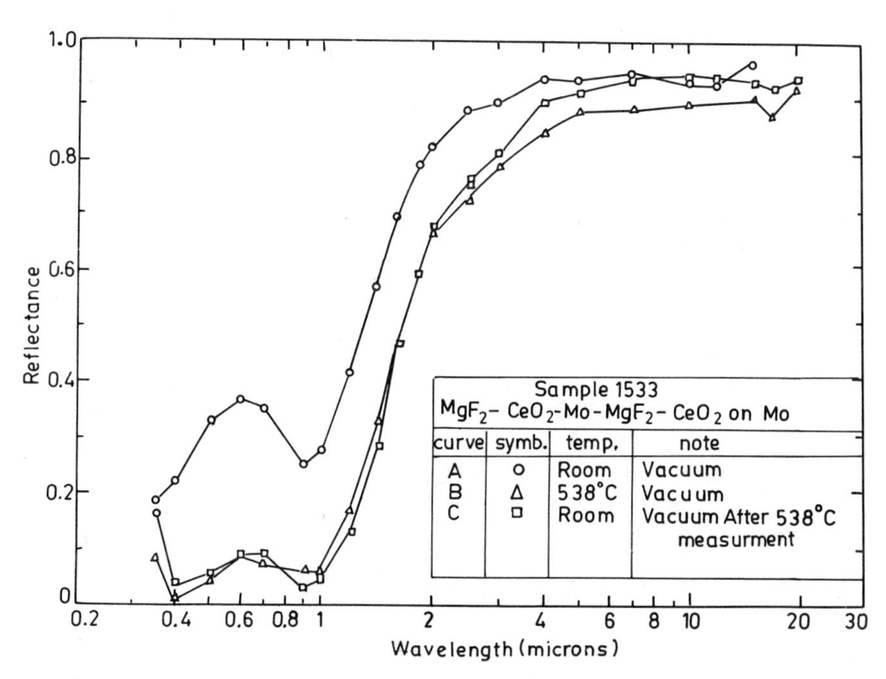

Fig. 5.8.8 Spectral reflectance of MgF_2–CeO_2–Mo–MgF_2–CeO_2 on Mo. Adapted from Schmidt and Park (1965) (Ref. 303).

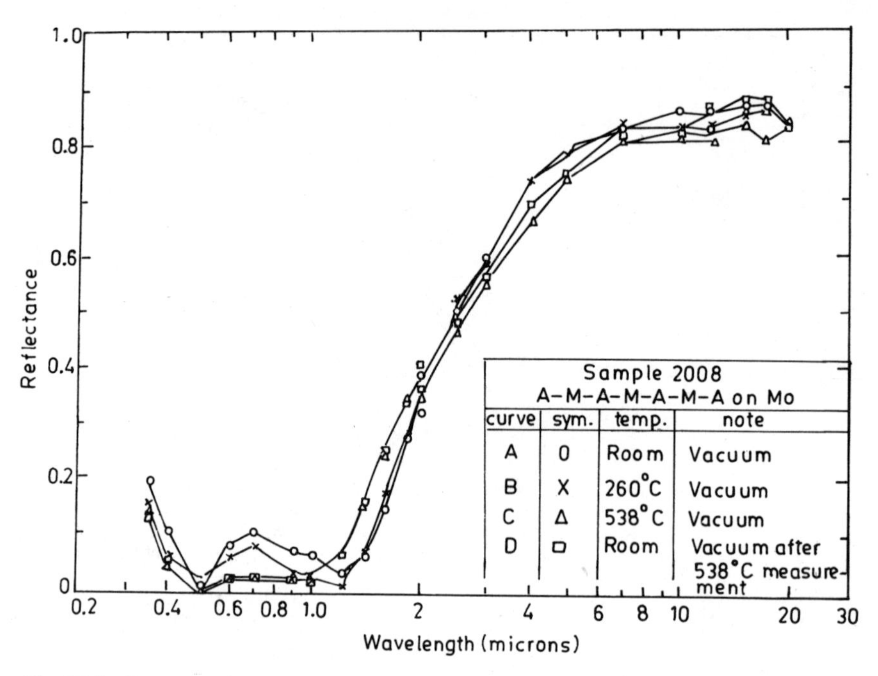

Fig. 5.8.9 Spectral reflectance of AMAMAMA on Mo. Adapted from Schmidt and Park (1965) (Ref. 303).

Table 5.8.1 Solar Absorptance and Thermal Emittance of Several Selective Coatings

Coating	Temperature (°C)	α_s	ε_T
Multilayer Stack			
Black Nickel[a]			
Black Nickel[b]			
Black Nickel[c]	100	0.95	0.07
MgF_2–Mo–CeO_2 on Mo[d]	260	0.85	0.053
	538	0.85	0.062
MgF_2–Mo–MgF_2–Mo–MgF_2 on Mo[d]	260	0.89	0.075
	538	0.89	0.090
MgF_2–CeO_2–Mo–MgF_2–CeO_2 on Mo[d]	260	0.85	0.073
	538	0.85	0.084
Al_2O_3–Mo–Al_2O_3–Mo–Al_2O_3–Mo–Al_2O_3[d] on Mo	260	0.91	0.085
	538	0.91	0.16
SiO_2–Mo–SiO_2 Mo[d]	Room temp.	0.86	0.08
CeO_2–Mo–CeO_2 on Mo[d]	Room temp.	0.90	0.06

[a] *Source:* Ref. 232.
[b] *Source*: Ref. 304.
[c] *Source:* Ref. 307.
[d] *Source:* Ref. 303.

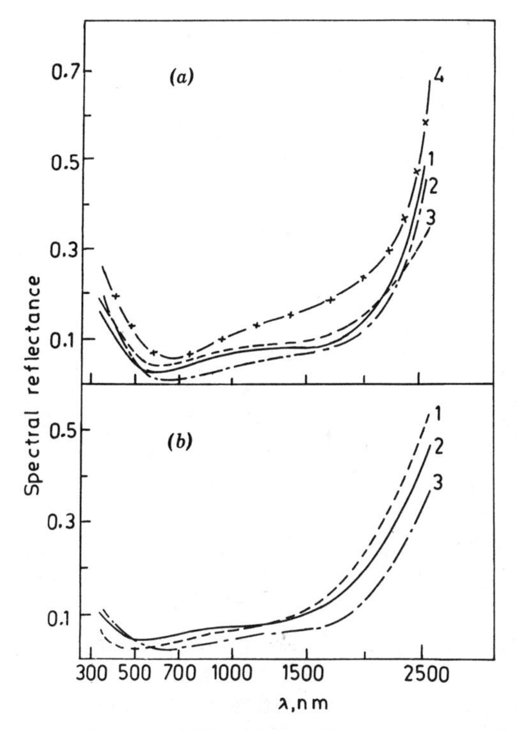

Fig. 5.8.10 (a) Spectral reflectance of SiO_2 – Mo – SiO_2 coatings deposited on (1) Mo foil ($\varepsilon = 0.08$); (2) on glass coated with opaque aluminium layer ($\varepsilon = 0.08$); (3) on steel ($\varepsilon = 0.15$); (4) Mo–SiO_2 coating on glass plus aluminium ($\varepsilon = 0.06$). (b) Spectral reflectances of SiO_2–Mo–SiO_2 coating deposited on glass plus aluminium as function of antireflective film thickness on upper and lower layer of SiO_2 (1) 730 Å; (2) 850 Å; and (3) 1000 Å. Adapted from Li et al. (1977) (Ref. 306).

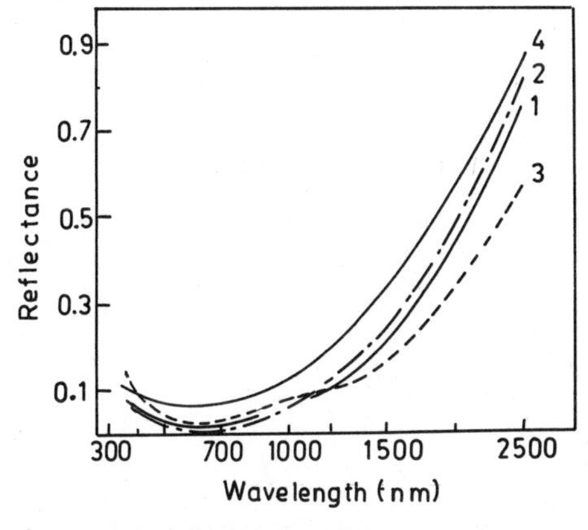

Fig. 5.8.11 Spectral reflectance of CeO_2–Mo–CeO_2 coating deposited on (1) Mo foil ($\varepsilon = 0.06$); (2) on glass plus aluminium ($\varepsilon = 0.07$); (3) on steel ($\varepsilon = 0.18$); (4) of coating Mo–CeO_2 ($\varepsilon = 0.06$). Adapted from Li et al. (1977) Ref. 306.

absorptance to high thermal reflectance can be shifted by depositing anti-reflective films of different thicknesses.

The disadvantage with most of the multilayer coatings is that they are quite expensive. They suffer from corrosion and interdiffusion at elevated temperatures which results degradation in optical properties. Some coatings are sensitive to abrasion. The AMA absorber is quite stable at elevated temperatures and does not suffer these drawbacks.

5.9 Optical Trapping Systems

Surface texturing is a common technique to obtain wavelength discrimination optical trapping of solar energy. Properly textured surfaces appear rough and absorbing to solar energy while appearing mirror-like and highly reflective to thermal energy. Tabor (209, 232) in 1957 proposed a method of enhancing the solar absorptance almost to unity by corrugating the surface into a series of V's. Further studies have been made by Montlouis Group (308, 309), Australian Group (310, 311), IBM Group (312, 313), and Seraphin (83) on these solar selective surfaces. Several authors tried to enhance solar absorptance of the surface by using surface texture such as wire mesh (314), grooves (310), electrodeposited coating on mechanically roughened surfaces (315), semiconductors evaporated in partial vacuum (242, 271), roughness produced by sputtering (316, 317), and CVD (312, 313), etc. Cuomo et al. (312) suggested that a microstructure similar in geometry to an acoustic anechoic surface may be used for optical trapping. The surface would consist of a dense forest of aligned needles whose diameters are of the order of visible wavelengths and the spacing between which is several wavelengths. This surface would absorb with high efficiency because of multiple reflections as the incident photons penetrate the needle maze.

A V corrugated surface is shown in Fig. 5.9.1. The radiation whose projection on the plane of the paper is normal to the folded surfaces suffers various interreflections. But as the projected angle of incidence increases,

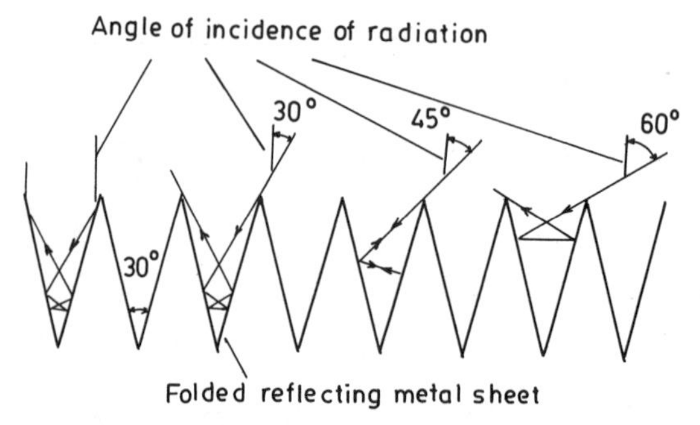

Fig. 5.9.1 Absorption of solar radiation due to successive reflections on V-corrugated surface. Adapted from Trombe et al. (1961) (Ref. 309).

incident radiation suffers lesser and lesser reflections (Fig. 5.9.1). The radiations whose projected angle of incidence is 90° suffers only one reflection. (Fig. 5.9.1). Holland (310) has described a method for determining the optimum angle of opening of the V corrugations for a particular application. If the surface is opaque, Kirchhoff's law gives $\alpha = 1 - r$. If the radiation suffers n reflections before emerging, its emergent intensity is r^n and the fraction absorbed is $1 - r^n$, i.e., the effective absorptance is

$$\alpha_e = 1 - r^n = 1 - (1 - \alpha)^n$$

or

$$\alpha_e = 1 - r^2 = 1 - (1 - \alpha)^2 \quad \text{for} \quad n = 2$$

If $\alpha = 0.8$, then $\alpha_e = 0.96$.

For low-emittance materials (207), the energy emitted from a V compared to that emitted from a flat plate is slightly less than A/a, where A is the true surface area of the V and a is the aperture of the V. Therefore for 60° V, the emittance will be slightly less than double the emittance of flat surface.

Recent investigations on the CVD of rhenium (83), tungsten (312), and nickel (318) dendrites on various substrates demonstrated the feasibility of above approach for producing solar selective absorbers. The dendrites appear as conical needles or whiskers. The dendrite forests absorb high-energy solar radiation by the geometry of multiple absorption and reflection as shown in Fig. 5.9.2. The sum of these absorptions is very high with respect

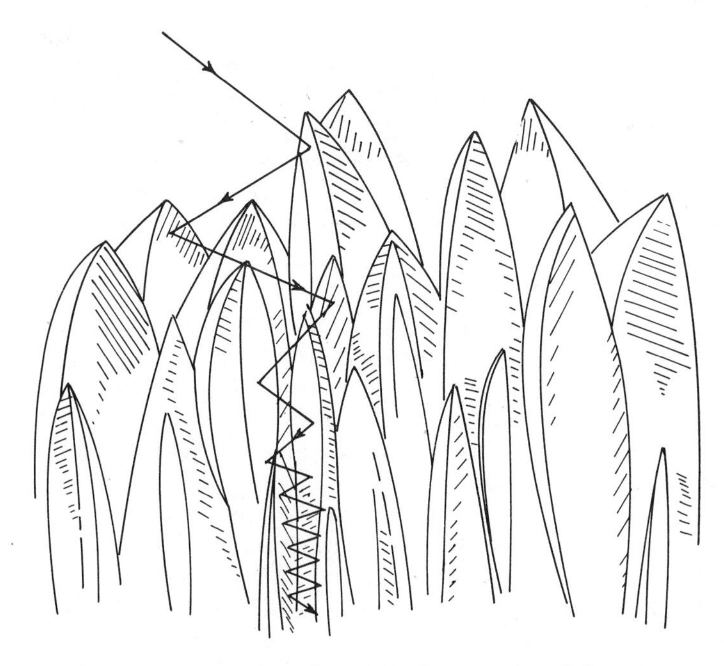

Fig. 5.9.2 Schematic representation of dendritic forests. Adapted from Lampert (1979) (Ref. 85).

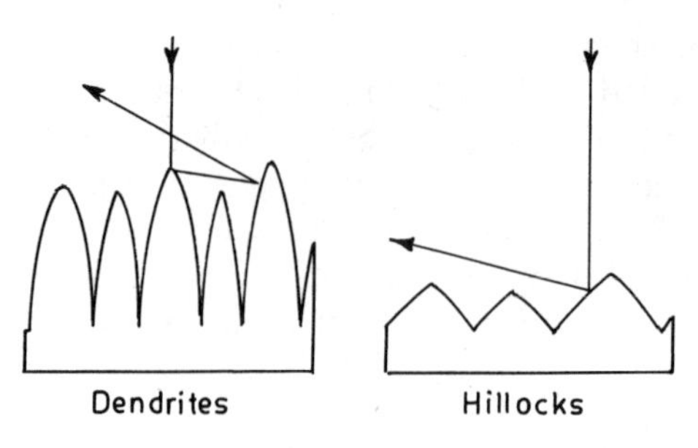

Fig. 5.9.3 Schematic representation and surface structure of a dendritic and hillock tungsten surface. Adapted from Pettit et al. (1978) (Ref. 313).

to the incoming radiation, although a single absorption may be low. In thermal region, the energy wavelengths are much larger than the dendrite spacings; the surface appears fairly smooth and acts like a highly reflecting or poor emitting surface. The dendritic tungsten surfaces were prepared by hydrogen reaction (319) of WF_6 at 500°C. Tungsten hexaflouride was mixed at 7–10% into a stream of purified hydrogen flowing at a rate of 10–12 liter/sec at atmospheric pressure. Tungsten films were deposited at 500°C on RF resistance-heated substrates, including sapphire, quartz, tungsten, stainless steel, carbon, and molybdenum. The structure of the deposited dendritic films was found to be a fiber texture with (111) planes oriented parallel to the surface of the substrate. The dendritic structure of the surface is shown in Fig. 5.9.3. For a thin layer of thickness from 2.5 to 25.0 μm, the surface is called a "gray dendritic" or "hillock" surface and for a thick film of thickness from 25 to 250 μm, the surface is called a "black dendrite" surface. The black dendritic surfaces have facet-included angle of less than 90°. For normal incidence illumination, the radiation is forced to undergo two or more reflections before re-emerging. For the gray dendritic surfaces, the facet angles are greater than 90°, the fiber heights are short, and the radiation undergoes two or less bounces before re-emerging. The anodization of the tungsten surfaces was performed in 0.1 N phosphoric acid (313). The thickness of the resulting tungsten oxide coating was determined by the limiting value of the anodization voltage chosen. Pettit et al. (315) reported that anodization of dendritic tungsten is found to have an advantage either in creating a solar absorber of extremely high solar absorptance when applied to a large dendritic surface or in enhancing the solar absorptance to thermal emittance ratio when applied to smaller hillock surfaces. The angular dependence of absorptance is reduced by the anodization coating which consists of WO_3. The directional dependence of solar absorptance for black and gray dendritic surface with and without an antireflective coating is shown in Fig. 5.9.4. The antireflective coating increases the solar absorptance from 0.76 to 0.90 and simultaneously decreases its dependence on incident angle. For hillock surfaces, the emittance value is 0.18 and the selectivity is 5.

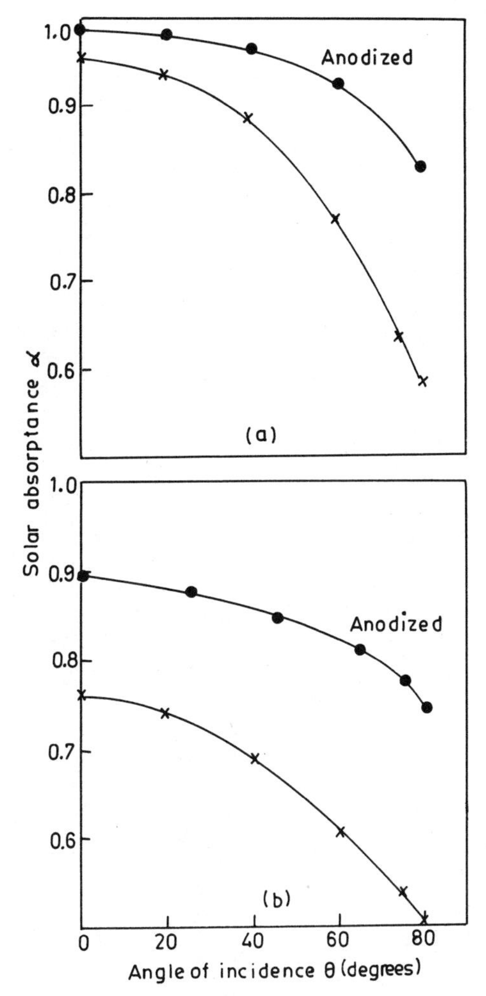

Fig. 5.9.4 Directional dependence of solar absorptance for (a) black dendritic and (b) hillock surfaces with and without antireflective coating. Adapted from Pettit et al. (1978) (Ref. 313).

Recently Grimmor et al. (318) investigated solar selective properties of Ni dendrites formed from CVD of nickel from Ni $(CO)_4$. Nickel dendritic absorber coatings are more attractive because of the low IR emissivity and low cost compared to tungsten dendrities. The experimental setup is described by Grimmor et al. for the preparation of nickel dendrites. Table 5.9.1 summarizes the solar absorptance and thermal emittance of several coatings having different thicknesses determined by scanning electron microscopy. Figure 5.9.5 shows the solar absorptance and thermal emittance properties of Ni dendrites as a function of thickness. Maximum value of solar absorptance α_s was found in 6 μm thick films, beyond which the value of α_s decreases. As the coating thickness increases from 0 to 20 μm, the emissivity increases

Table 5.9.1 Experimental Results for the Formation of Ni Dendrites from the CVD of Ni from $Ni(CO)_4$

Run	SEM Observed Coating Thickness	SEM Dendrite Spacing (W) and Appearance	ε_T	α_s	Physical Appearance to Unaided Eye
8–7–75	6×10^{-4}cm	$\geq 1.7 \times 10^{-4}$cm rounded spikes	0.599[a] (0.71)[c] (0.89 at 10 μm)[e]	0.950[b] (0.94)[d]	Very black
8–13–75	2.2×10^{-3}cm	$\geq 1.7 \times 10^{-4}$cm rounded spikes and buds	0.3 99[a] (0.66)[c] (0.64 at 10 μm)[e] (0.75 at 5 μm)[e]	0.878[b] (0.94)[d]	Light gray, poor adhesion (spotty)
8–18–75	1×10^{-3}cm	$\geq 1.7 \times 10^{-4}$cm sharp spikes	0.710[a] (0.79)[c] (0.98 at 10 μm)[e] (0.99 at 5 μm)[e]	0.931[b] (0.94)[d]	Dark gray
9–3–75	1×10^{-2}cm	W_{small} $\geq 2 \times 10^{-4}$cm W_{large} $\geq 4 \times 10^{-3}$cm flowery spikes	0.558[a] (0.89)[c] (0.86 at 10 μm)[e]	0.893[b] (0.92)[d]	Medium gray
10–6–75	2×10^{-4}cm	$\geq 1.3 \times 10^{-4}$cm rounded buds	0.20[f] (0.47 at 10 μm)[e] (0.46 at 5 μm)[e]	0.899[f]	Medium gray (thin coating)
10–21–75	2×10^{-3}cm	$\geq 2.7 \times 10^{-4}$cm long, sharp, curving spikes	0.747–0.711[a] (0.95 at 10 μm)[e] (0.91 at 5 μm)[e]	0.954–0.931[b]	Dark gray to black

Source: Adapted from Grimmer et al. (1978) (ref. 318).
[a] Value of $\varepsilon_{T,H}$ (100°C) obtained by Pettit of Sandia Labs., Albuquerque, NM, using a Gier–Dunkle DB-100 instrument.
[b] Value of α_s obtained by Pettit of Sandia Labs., Albuquerque, NM, using a Gier–Dunkle MS-251 instrument.
[c] Value of ε_T (100°C) derived from spectral data obtained using Backman IR 20A–X spectroreflectometer.
[d] Value of α_s derived from spectral data obtained by Myers, Sandia Labs., using Beckman DK-2 spectrophotometer with an integrating sphere.
[e] Value of $\varepsilon_{5\mu m}$ (100°C) and $\varepsilon_{10\mu m}$ (100°C) obtained by Carlos, using IR television cameras AGA 750 and AGA 680, respectively.
[f] The α_s value obtained using a Gier–Dunkle MS-251 instrument at LASL; the ε_{TH} (100°C) obtained using a Gier–Dunkle DB-100 instrument LASL.
[g] After polishing with crocus cloth $\varepsilon_{5\mu m}$ (100°C) = 0.56, $\varepsilon_{10\mu m}$ (100°C) = 0.65; bare aluminium $\varepsilon_{5\mu m}$ (100°C) = 0.29.

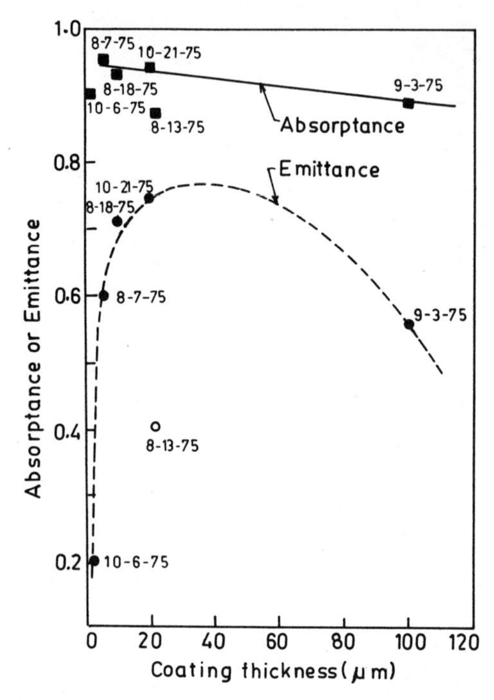

Fig. 5.9.5 Solar absorptance and thermal emittance of nickel dendrites as a function of thickness. Adapted from Grimmer et al. (1978) (Ref. 318).

Table 5.9.2 Solar Absorptance and Thermal Emittance of Optical Trapping Surfaces

Type		α_s	ε_T	α/ε	Structure
1	Tungsten dendrite[a]	0.99	0.26–0.3 (550°C)	3.8–3.3	Aligned dendrites
2	Nickel dendrite[b]	0.95	0.6 (100°C) 0.71 (100°C)	1.6 1.2	Aligned dendrites
3	Aluminium steel[c]	0.99	0.55 (260°C)	1.8	Random dendrites
4	Aluminium–Stainless Steel[a]	0·89	0.47 (260°C)	1.8	Random dendrites
5	Aluminium Nickel[c]	0.94	0.30 (260°C)	3.1	Porous cavities

Source: a—Ref. 313; b—Ref. 318; c—Ref. 320.

from ≈ 0.1 to a values of 0.75. The maximum value of emissivity is expected to occur in the 20 to 50 μm range. The curvature of the dendrite and subsequent surface bridging and melding becomes visible at IR wavelengths and leads to an increase in the reflectivity at these long wavelengths. Santala and Sabol (320) produced a dendrite coating directly on collector plates by reacting laminated metal layers in an exothermic atmosphere. The most successful of this type of intermetallic absorber is the aluminium–nickel surface, which consists of visible wavelength size cavities. Solar selective properties of various optical trapping surfaces are listed in Table 5.9.2. The reduction in size of surface roughening to the order of the wavelength of energy required to be absorbed can increase the net absorptance of the absorber, but simultaneously it may increase the thermal emittance. The amount of increase in emittance is related to the microroughness and crystallite size. The net effect of microroughness can be quite complex for a real absorber. At very fine surface roughness the reflective scattering mechanism, due to cavities, changes over to resonant scattering (discussed in the next section), which depends upon the material properties and morphology in the fractional micron size range.

5.10 Composite Materials Coatings

Composite films of small metal particles embedded in dielectric, also known as granular or cermet films, have optical properties appropriate for good selective solar absorbers. These films absorb strongly in the solar region due to interband transitions in the metal and the small particle resonance, while they are transparent in the thermal IR region. When cermet films are deposited on a highly reflecting metal mirror, the tandem forms a selective surface having a high solar absorptance and low thermal emittance. Resonant scattering depends upon both the size and optical properties of the particles and surrounding medium. The transition between reflective and resonant scattering separates the gold black deposits and the gold smoke filters (321, 322). These two deposits are similar in their physical structure, particle size, and degree of aggregation. Both the coatings absorb strongly the visible and near IR. Gold black deposit consists of submicron particles of pure gold deposited on insulating substrates, and retain the metallic properties of the bulk materials. Gold smoke deposits are evaporated under conditions that add tungsten oxide to a similar aggregate, rendering it an insulator.

The gold black absorbs strongly in the IR through the action of free metal electrons whereas gold smoke is transparent to IR. Therefore the small fraction of tungsten oxide in gold smoke particles gives rise to a drastic change in the optical properties of the deposits. McKenzie (323–325) produced gold black deposits by gold evaporation from a hot tungsten source in a nitrogen atmosphere and in a mixture of oxygen and nitrogen. Figure 5.10.1 shows specular reflectance of several gold black and gold smoke

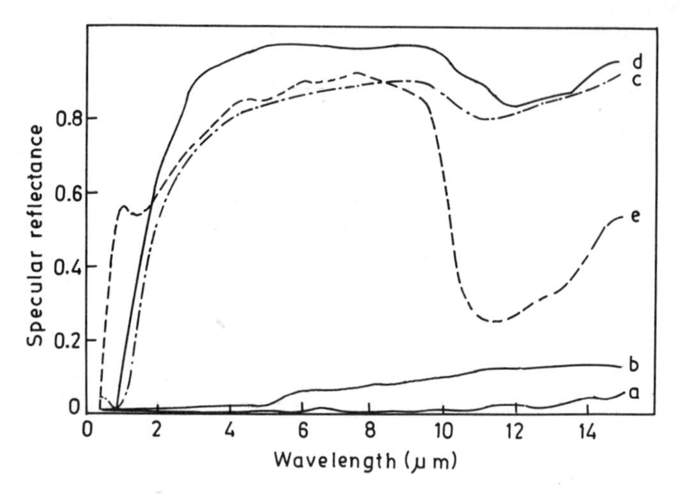

Fig. 5.10.1 Spectral reflectance of several gold black and gold smoke films deposited on copper under various conditions. (a) Gold black deposited in oxygen-free nitrogen atmosphere at 400 Pa; mass thickness 0.81 g/m^2. (b) Gold black deposited in nitrogen containing 5% oxygen at 400 Pa; mass thickness 0.90 g/m^2. (c) Gold smoke deposited in nitrogen containing 20% oxygen at 400 Pa; mass thickness 0.98 g/m^2. (d) Gold smoke deposited in nitrogen containing 20% oxygen at 1300 Pa; mass thickness 0.98 g/m^2. (e) tungsten oxide deposited in pure oxygen at 400 Pa; Mass thickness 0.97 g/m^2. Adapted from McKenzie (1978) (Ref. 323).

coatings deposited on copper substrate. The gold black reflectance curve shown in Fig. 5.10.1 for a sample of thickness 0.81 g/m^2 deposited from 10 torr air onto copper gives a value of solar absorptance $\alpha_s = 0.88$ and a thermal emittance $\varepsilon_T = 0.10$ at 400°C. The theoretical curve calculated from Mie's theory, for 240 nm gold particles of mass thickness 0.89 g/m^2 on copper gives values of solar absorptance $\alpha = 0.75$ and thermal emittance $\varepsilon = 0.07$ at 400°C.

Several authors (326–345) have done extensive investigations on the study of metal–insulator and semiconductor–insulator composites produced by cosputtering, electroplating, and various forms of vapor deposition. Electroplated chrome black consists of a graded composite of chromium and chromium oxide, and is one of the most widely used selective surface (83, 84, 326, 328).

The black chrome coatings produced by electroplating process have an advantage of low cost but these are not stable at elevated temperatures (> 300°C). Recently many researchers have paid attention to the vapor deposited composite films which have good stability at high temperatures and simultaneously have high solar absorptance and low thermal emittance.

5.10.1 *Metal–Insulator Composite Films*

The dispersion of metal particles in a dielectric or conductive host matrix provides spectral selectivity through resonance scattering phenomenon.

A dispersion of vanadium, calcium, and niobium metal in copper metal exhibits a broad resonance in the solar region and low emittance in the thermal region. The resonant frequency of small particles of transition metals in the dielectric composite structure may be adjusted to occur at the solar maximum by controlling optical properties and density of the particles and the medium in which they were embedded (329). There have been numerous investigations of the optical properties of gold and silver particles (330, 342–346) dispersed in dielectrics or in island film structures. The theoretical description of the optical properties of composite materials was given by Garnett in 1904 (349). The same result in detail was derived by Lamb et al. (348). In particular, two effective medium theories have been used to interpret the experimental data, namely, the non-self-consistent Maxwell–Garnett approach (249), and the self-consistent Bruggeman theory (350, 351). The two theories are described in brief in the next section.

Fan and Zavracky (343) and Gittleman et al. (341) prepared fine grained Au/MgO films by cosputtering from a hot-pressed 12.7 cm diameter target containing 75 vol. % MgO and 25 vol. % Au. Figure 5.10.2 shows extinction coefficient *k* and refractive index *n* as a function of wavelength for a 1500 Å Au–MgO composite cermet film reported by Fan and Zavracky (343) and

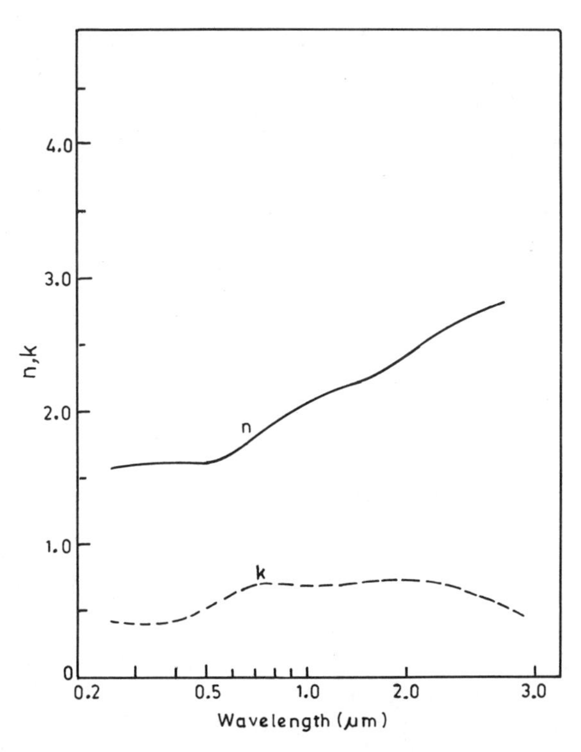

Fig. 5.10.2 Wavelength dependence of refractive index *n* and extinction coefficient *k* for Au–MgO films. Adapted from Fan and Zavracky (1977) (Ref. 343).

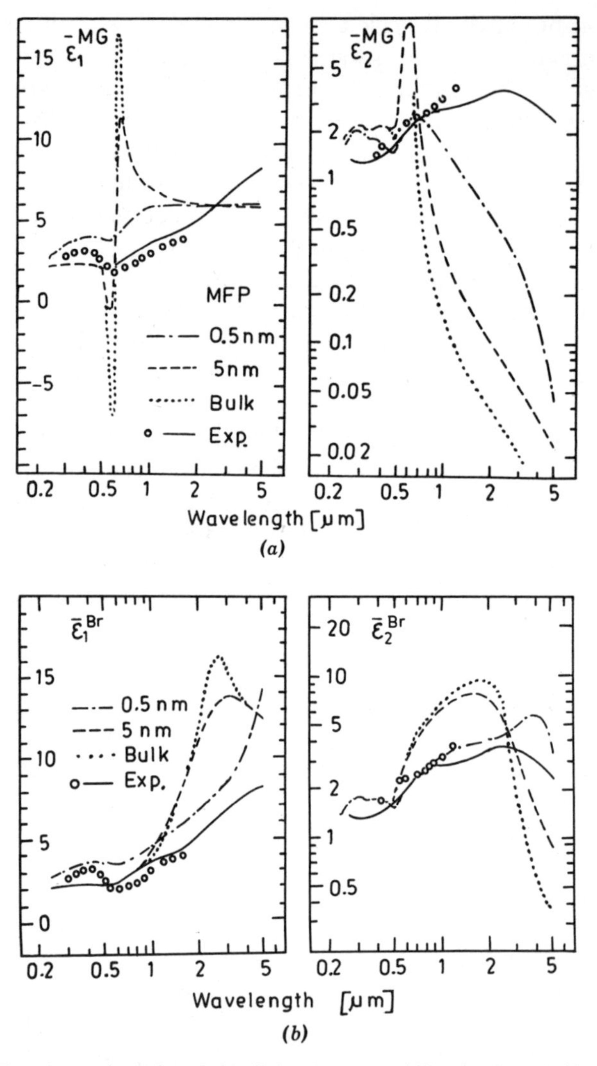

Fig. 5.10.3 Experimental results of the dielectric permeability for 25 vol. % Au–MgO are shown by solid curves [Fan and Zavracky (343)], and by circles [Fan and Spura (293) and Gitteman et al. (341)]. Dashed and dotted curves based on (a) Maxwell–Garnett and (b) Bruggeman theory, with three different values of the electron mean free path (MFP). Adapted from Granqvist (1979) (Ref. 352).

Gittleman et al. (341). In Fig. 5.10.3, the experimental values of the real part ($\varepsilon_1 = n^2 - k^2$) and imaginary part ($\varepsilon_2 = 2nk$) of dielectric permeability are combined with the values calculated by Granqvist (352). The dashed and dotted curves denote computations pertaining to the Maxwell–Garnet theory and to the Bruggeman theory, respectively. The dotted curves were obtained by using the bulk mean free path of Au in the calculations. The figure shows that both theories disagree with experimental values. In the beginning, Granqvist (352) assumed that the mean free path is equal to the particle radius which is of the order of 10 nm for Au–MgO cermet films, i.e. the mean free path $l = 5$ nm should be a good approximation to account for boundary scattering. It is seen in Fig. 5.10.3 that this approach is in agreement with the experimental results. The gold cermet films are generally produced under strongly off-equilibrium conditions. This causes a large amount of various imperfections in gold particles which in turn strongly depresses the mean free path. Granqvist showed that the Bruggeman theory gives good agreement with experiments when a very small value of mean free path ($l = 0.5$ nm) is used.

The solar absorptance $\alpha_s = 0.90$ to 0.93 and thermal emittance $\varepsilon_T = 0.04$ for copper substrate and 0.1 for stainless-steel substrate were reported by Fan and Zavracky (343) for Au–MgO cermet films prepared by RF sputtering. These films are stable up to 400°C when deposited on stainless steel coated with 1000 Å thick molybdenum. Figure 5.10.4 shows a comparison of spectral reflectivities of as-grown and heat-treated Au–MgO cermet films. The heat treatment was carried out at 400°C for 64 hr (343). Degradation began to occur when samples were heated up to 500°C, which causes a decrease in solar absorptance without changing thermal emittance.

Granqvist extended the above analysis to the 18 vol. % Au + Al_2O_3 cermet films produced by coevaporation which were recently studied by McKenzie and McPhedran (353). Figure 5.10.5 shows a comparison between experimental values and the Maxwell–Garnett and Bruggeman theories. The calculations were made by taking the same values of mean free path as in Au–MgO films. It was observed that the experimental results agree well with Bruggeman's theory.

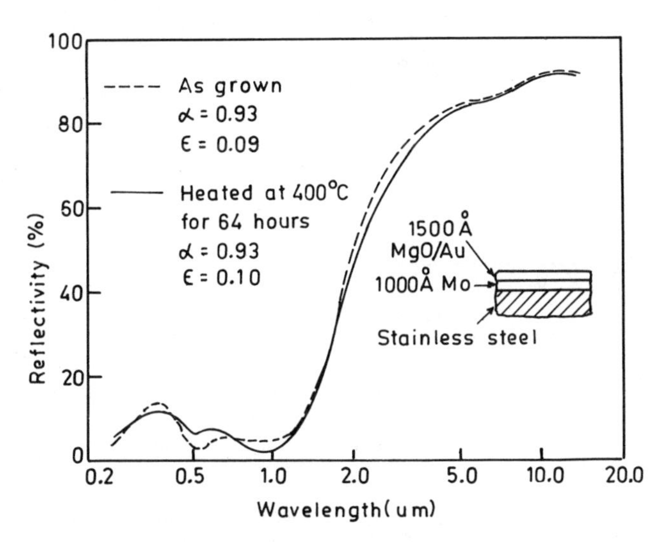

Fig. 5.10.4 Spectral reflectance of an as-grown and a heat-treated Au–MgO films deposited on molybdenum coated stainless steel. Adapted from Fan and Zavracky (1977) (Ref. 343).

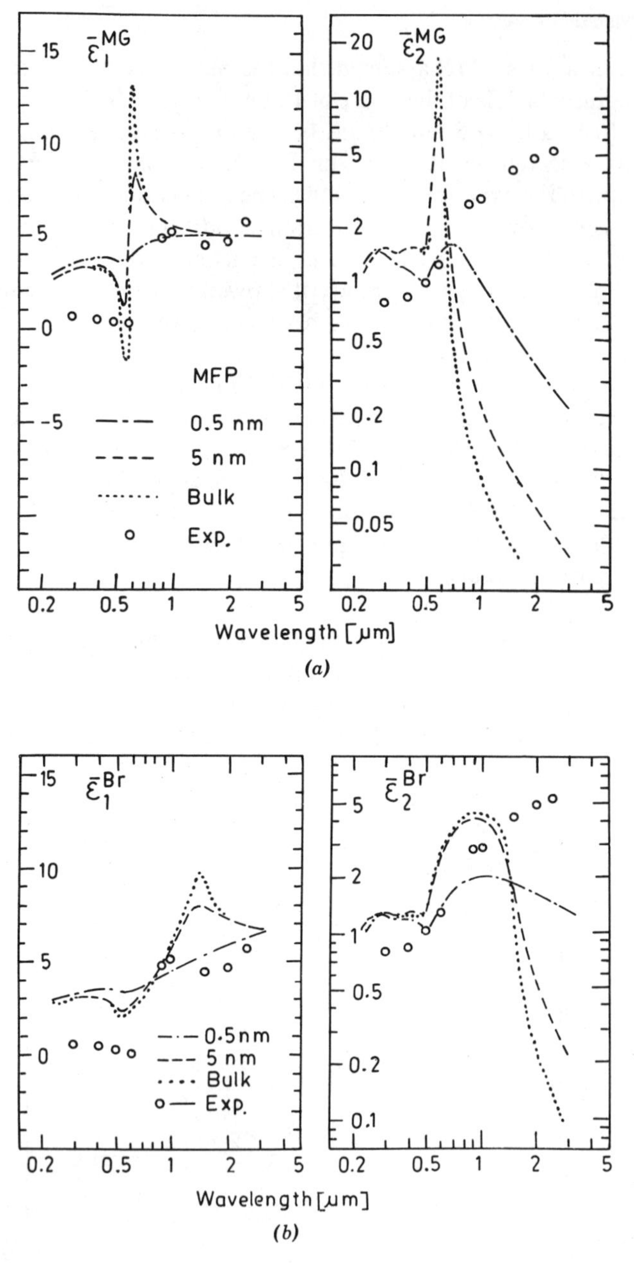

Fig. 5.10.5 Experimental results of dielectric permeability for 18 vol. % $Au-Al_2O_3$ (circles) Dashed and dotted curves show the calculations based on (a) Maxwell–Garnett theory and (b) Bruggeman theory with three different values of electron mean free path (MFP). Adapted from Granqvist (1979) (Ref. 352).

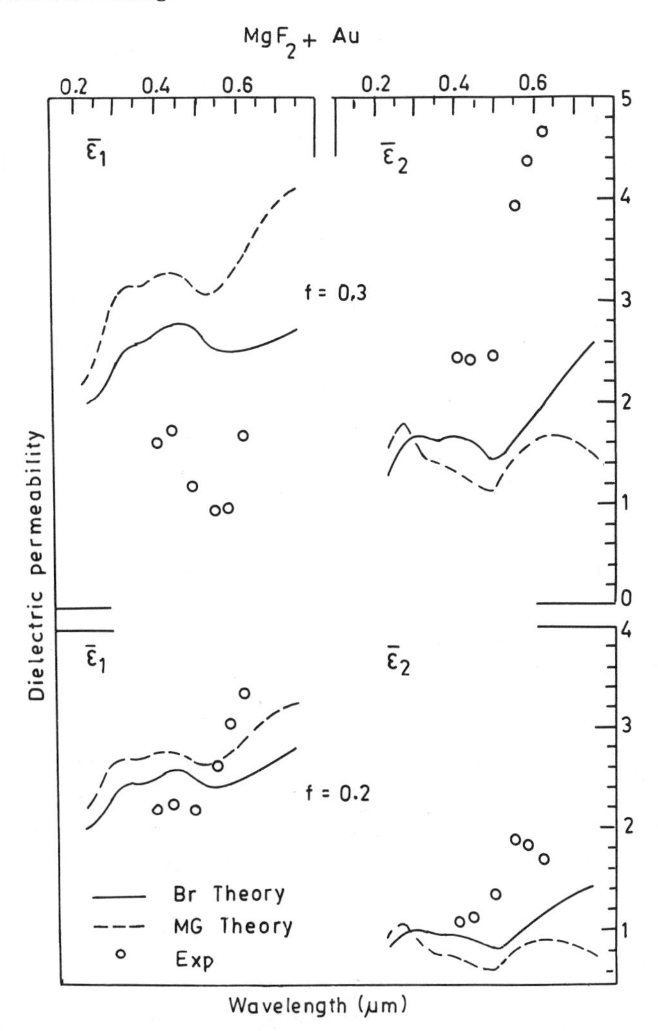

Fig. 5.10.6 Experimental results of dielectric permeability for 20 vol. % Au–MgF_2 and 30 vol% Au/MgF_2 Dashed and dotted curves show calculations based on Maxwell–Garnett theory and Bruggeman theory with an electron mean free path of 0.5 nm. Adapted from Lissberger (1974) (Ref. 354).

In case of Au–MgF_2 cermet films, experimental results were reported by Lissberger and Nelson (354). Granqvist (352) concluded that the Maxwell–Garnett theory was in reasonably good agreement with the measurements. The filled circles in Fig. 5.10.6 denote the experimental data for cermet films with 20 vol. % Au and with 30 vol. % Au. The curves show results of computations where Granqvist used $\varepsilon_m = 1.9$ together with mean free path $l = 0.5$ nm.

Considerable attention has been paid to Cr–Cr_2O_3 cermet films in recent years by various authors (84, 220, 293, 355). Fan and Spura (293)

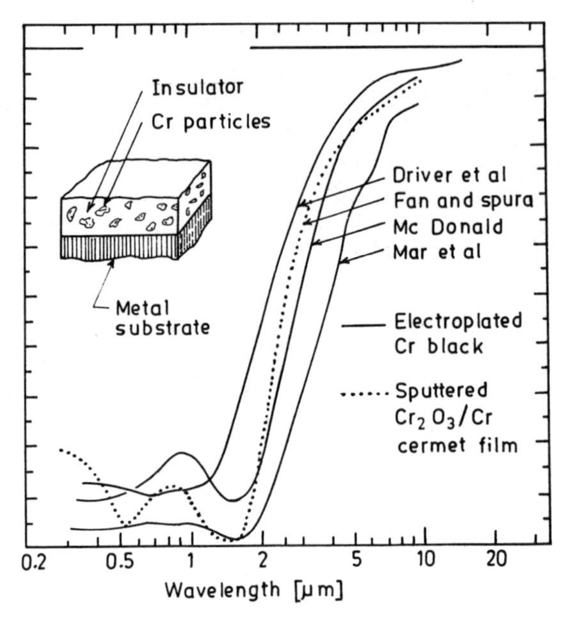

Fig. 5.10.7 Spectral reflectance of chromium black deposited by different techniques. The inset shows Granqvist's conceptual model for chromium black. Adapted from Granqvist (1979) (Ref. 356).

studied optical properties of Cr–Cr_2O_3 cermet films produced by an RF sputtering technique. The cermet films were deposited on metal substrates and overcoated with a Cr_2O_3 antireflection coating. Figure 5.10.7 shows a comparison of spectral reflectance of spectrally selective black chrome prepared by different techniques. Granqvist and Hunderi (356, 357) computed spectral reflectance from metal surfaces with coatings of ultrafine metal particles dispersed in insulating materials. They also analyzed the role of coating thickness, substrate metal, particle shape and orientation, possible dielectric cores in particles, dielectric permeability of the embedding medium, and graded volume fractions of metals.

The composite cermet film of 29 vol. % Cr and an antireflection coating has excellent selective properties with solar absorptance of about 0.92 and thermal emittance of about 0.08. The films were stable and no changes in solar selective properties are observed after heat treatment at 300°C for 60 hr. At 400°C, the heat treatment caused a crack and in some areas the substrate peels off, probably because of difference in thermal expansion coefficients (293).

In recent years, Craighead and Buhrman (358, 359) and Craighead et al. (360) investigated solar selective properties of composite films of Ni, V, Fe, and Pt particles embedded in Al_2O_3, MgO, and SiO_2 matrices; the films were produced by controlled coevaporation. The materials were evaporated from two independently controlled electron-beam evaporators in an oil-free vacuum system with typical system pressures before evaporation of 3×10^{-8} torr. The substrates were either fused quartz or polished copper and could be heated as high as 500°C. The microstructure of the evaporated Ni–Al_2O_3 and Pt–Al_2O_3 composite films were studied by electron microscopy and electron diffraction techniques. The Ni–Al_2O_3 composite film formed a structure of crystalline Ni particles embedded in an amorphous matrix. For small metallic volume fraction $F \leq 0.2$, the Ni particles were nearly spherical typically with diameters of 5–10 nm. The particle size depended on substrate temperature during deposition, with particles growing to larger size on heated substrates. For larger Ni volume fractions, the particle shapes were more distorted. For Pt–Al_2O_3 composite films, electron microscopy revealed no resolvable particles, i.e. no particles larger than 2 nm. The electron diffraction pattern showed only diffuse ring which is a characteristic of amorphous solids (360).

For V and Fe composite films, no metallic particles were observed by electron microscopy. The MgO in the Fe–MgO and V–MgO system forms crystals but no electron diffraction pattern could be seen from the metal. The V–SiO_2, V–Al_2O_3, and Fe–Al_2O_3 composite films show no crystalline diffraction pattern. The observed structure of the V and Fe composites was independent of substrate deposition temperature from the room temperature to 500°C. The metal in the V and Fe composites is either in the form of very small particles or uniformly dispersed throughout the material.

Considerations of the relative thermodynamic stability of the oxides of Ni, Fe, and V compared to the host oxide Al_2O_3, MgO, and SiO_2 explain that it is difficult to form Fe and V composites with the simple structures corresponding to the model of Maxwell–Garnett theory. Craighead and Buhrman (358) concluded that V and Fe composites have optical properties different in character from the predictions of Maxwell–Garnett theory and are not as suitable for solar selective surfaces as Ni composites.

The total hemispherical reflectance is shown in Figs. 5.10.8 and 5.10.9 for three Ni–Al_2O_3 and Pt–Al_2O_3 composite films reported by Craighead et al. (360). The major difference of the films in each family is the film thickness. In each case the thinnest film is about 10 nm thick and the thickest about 600 nm. The total hemispherical emissivity of all the composite films deposited on copper, measured by calorimetric technique, is shown in Fig. 5.10.10 and 5.10.11. Table 5.10.1 lists the solar absorptance and thermal emittance of Ni and Pt composite films of different thicknesses. The solar performance of composite films is shown in Figs. 5.10.12 and 5.10.13, using the measured value of solar absorptance and thermal emittance (360) at 150 and 500°C. Craighead (360) reported that Ni and Pt composite films deposited on copper were able to survive several small heatings to above 600°C in vacuum. A Ni–Al_2O_3 composite film was heated in air at 300°C for 50 hr. The spectral reflectance is shown in Fig. 5.10.14 for as-grown and heat-treated samples. Figure 5.10.15 shows spectral reflectance curves for a Pt–Al_2O_3 composite film deposited on copper for as-grown and samples heated at 300°C for

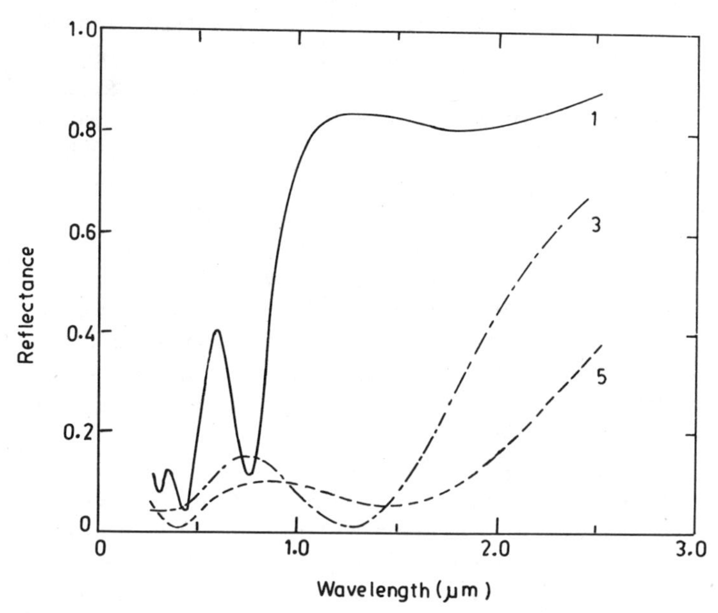

Fig. 5.10.8 Spectral reflectance of three Ni–Al_2O_3 composite films having different thicknesses deposited onto polished copper. Curve 1 corresponds to a film having a thickness of 100 nm and curve 5 corresponds to 600 nm thickness. Curve 3 corresponds to film of intermediate thickness. Adapted from Craighead et al. (1979) (Ref. 360).

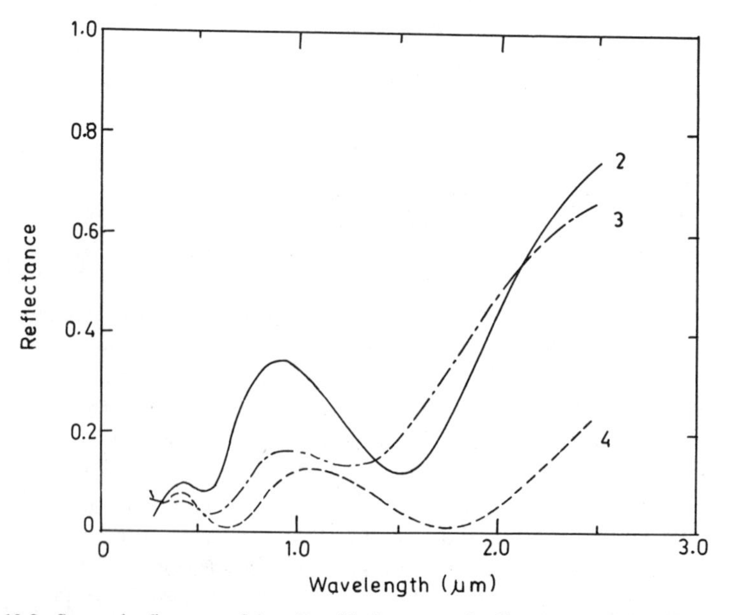

Fig. 5.10.9 Spectral reflectance of three Pt–Al_2O_3 composite films having different thicknesses deposited onto polished copper. Curve 2 corresponds to 100 nm thick film, curve 4 corresponds to 600 nm thick film, and curve 3 corresponds to film of intermediate thickness. Adapted from Craighead et al. (1979) (Ref. 360).

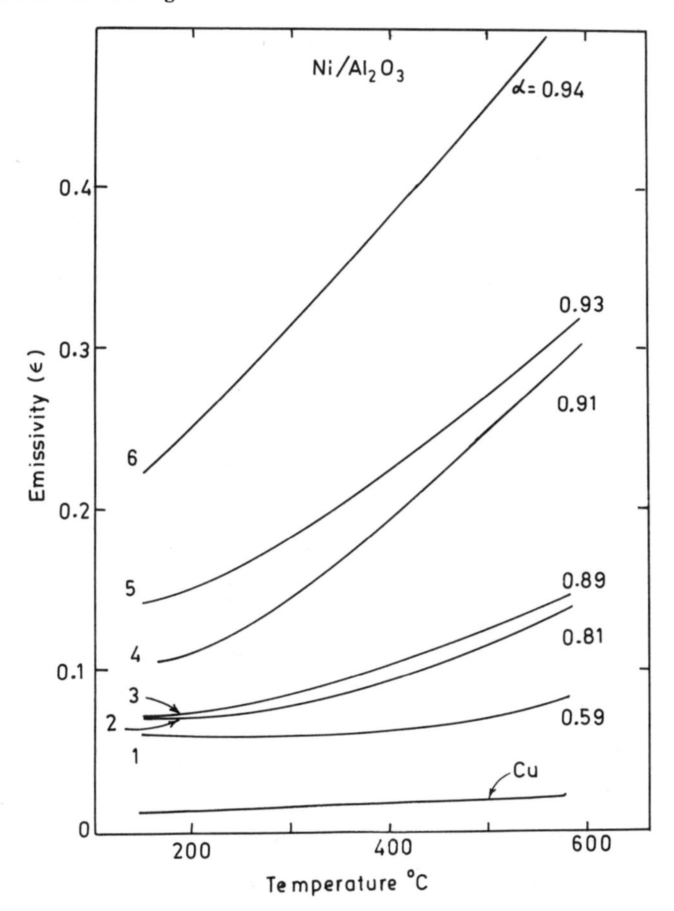

Fig. 5.10.10 Spectral emittance of Ni–Al_2O_3 composite films deposited on copper and bare copper substrate. Adapted from Craighead et al. (1979) (Ref. 360).

50 hr and then at 400°C for 50 hr. No appreciable change was observed after heat treatment at 300 and 400°C, but after treatment at 500°C the Pt–Al_2O_3 film flaked off the copper substrate. A Pt–Al_2O_3 film deposited on a Pt film coated quartz substrate has shown no degradation in performance after heat treatment at 600°C for 300 hr in air.

5.10.2 *Semiconductor–Insulator Composite Films*

Gittleman et al. (361, 362) have studied composite semiconductors produced by cosputtering CaF_2 and MgO with either germanium or with silicon. The optical behavior of these composites can be described as being similar to that of the parent semiconductor with the same energy gap but with a reduced concentration-dependent refractive index. In silicon composite films, strong absorption bands appear in the IR. These are due to compounds

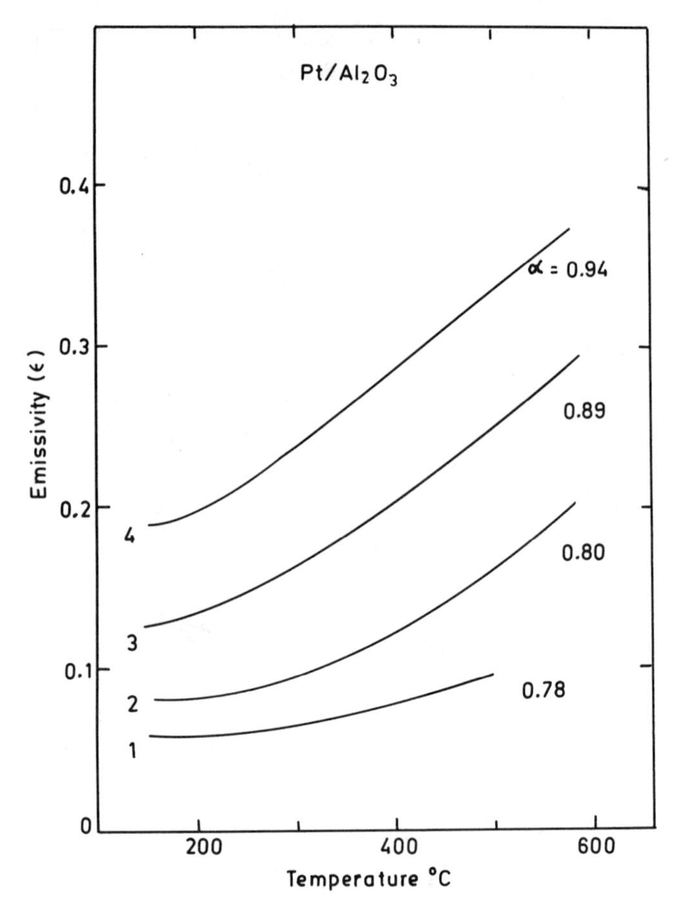

Fig. 5.10.11 Spectral emittance of Pt–Al_2O_3 composite films deposited on copper. Adapted from Craighead et al. (1979) (Ref. 360).

formed by chemical reaction between semiconductor and insulator matrix. The cermet films (or composite films) were deposited on appropriate substrate by RF sputtering as described by Abeles (363). Auger and ESCA studies for Si–CaF_2 showed that bound oxygen exists in a surface layer which is almost 100 Å thick. Beneath the surface layer there was no oxygen detectable and the silicon present was indistinguishable from bulk silicon. The refractive index n and extinction coefficient k of the composite semiconductor film were determined from transmittance and reflectance data (363). Figure 5.10.16 shows the refractive index n versus volume fraction F for Ge composite film. For F between 0.3 and 0.5, $n \approx 2$. The normal reflectance for two Ge–CaF_2 composite films having different thickness is shown in Fig. 5.10.17 (362). Both the films were deposited on substrates having a 1500 Å layer of evaporated aluminium. Similarly, Fig. 5.10.18 shows normal reflectance for Si–CaF_2 composite film 5 μm thick on an aluminium mirror.

Table 5.10.1 Solar Absorptance and Thermal Emittance of Ni–Al_2O_3 and Pt–Al_2O_3 Deposited on Copper

Sample[a]	α_s	ε_T(150°C)	ε_T(300°C)	ε_T(500°C)
Ni–Al_2O_3				
1	0.59	0.06	0.06	0.07
2	0.81	0.07	0.08	0.12
3	0.89	0.07	0.09	0.13
4	0.91	0.10	0.15	0.26
Pt–Al_2O_3				
1	0.78	0.06	0.07	0.09
2	0.80	0.08	0.10	0.16
3	0.89	0.12	0.16	0.25
4	0.94	0.19	0.24	0.33

Source: Adapted from Craighead et al. (1979) (Ref. 360).
[a]The sample label numbers are in order of increasing thickness

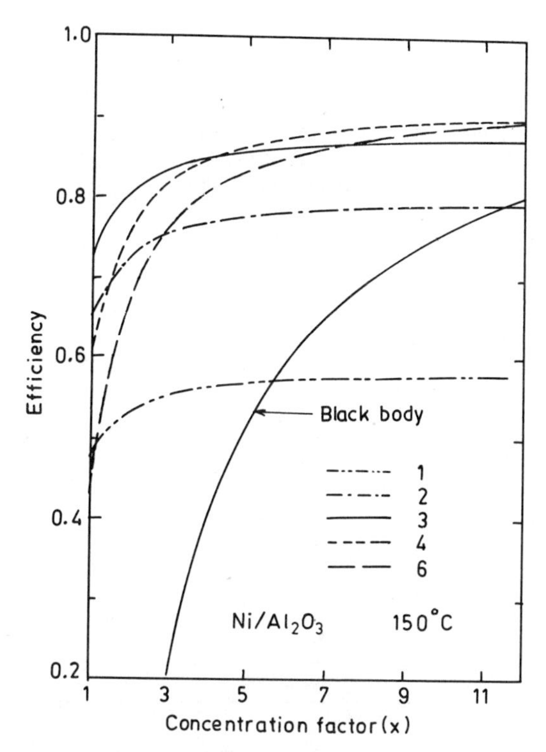

Fig. 5.10.12 The collection efficiency (in %) versus solar concentration for Ni–Al_2O_3 films at 150°C and their comparison with ideal blackbody. Adapted from Craighead et al. (1979) (Ref. 360).

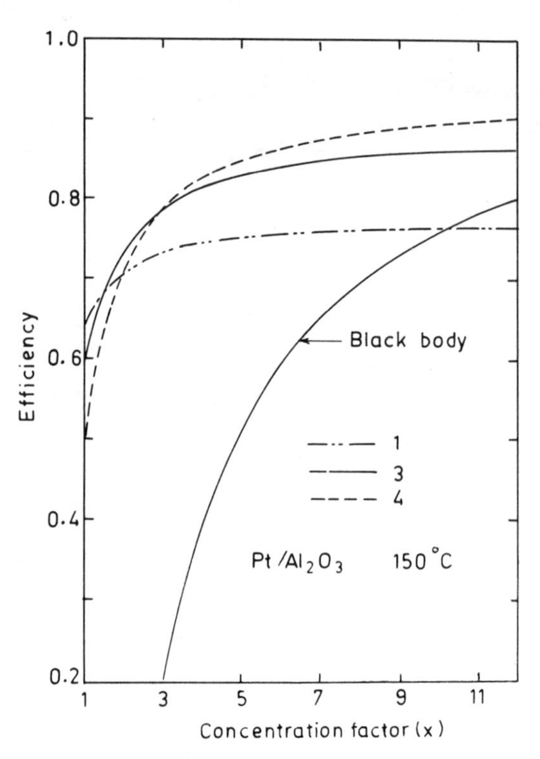

Fig. 5.10.13 The collection efficiency versus solar concentration for Pt–Al_2O_3 films at 150°C and their comparison with ideal blackbody. Curve 1 corresponds to a film having a thickness of 100 nm, curve 4 corresponds to a film having a thickness of 500 nm, and curve 3 corresponds to a film of intermediate thickness. Adapted from Craighead et al. (1979) (Ref. 360).

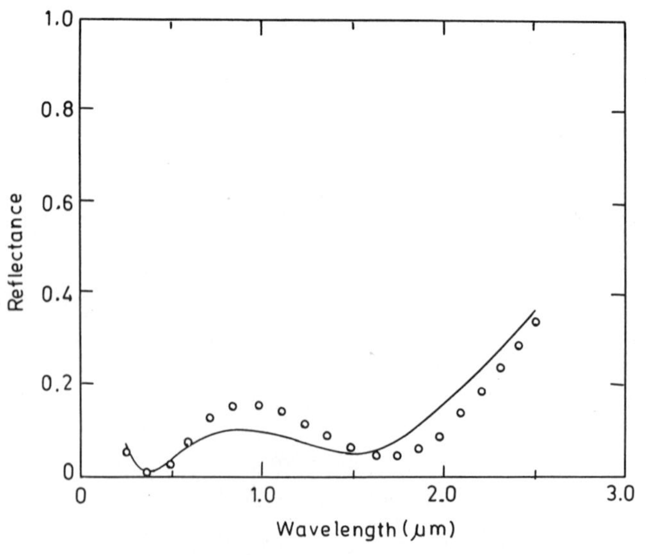

Fig. 5.10.14 Spectral reflectance of Ni–Al_2O_3 as-grown (solid lines) and heat treated (circles) at 300°C for 50 hr films. Adapted from Craighead et al. (1979) (Ref. 306).

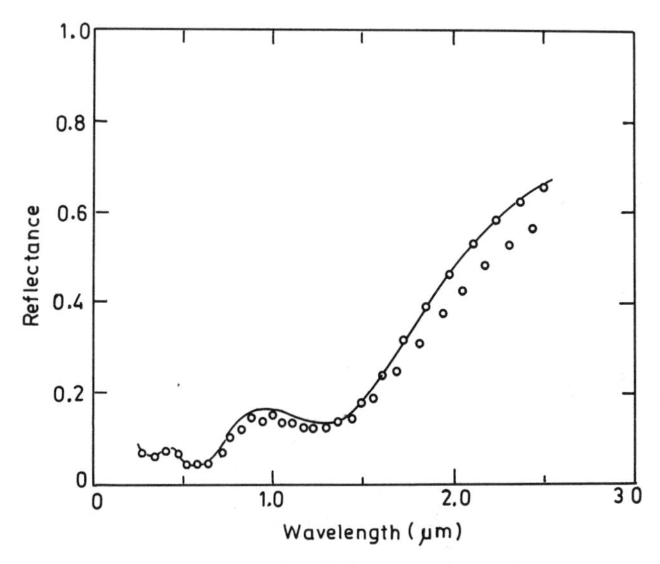

Fig. 5.10.15 Spectral reflectance of Pt–Al_2O_3 as-grown (solid lines) and heat treated (circles) at 300°C for 50 hr films. Adapted from Craighead et al. (1979) (Ref. 360).

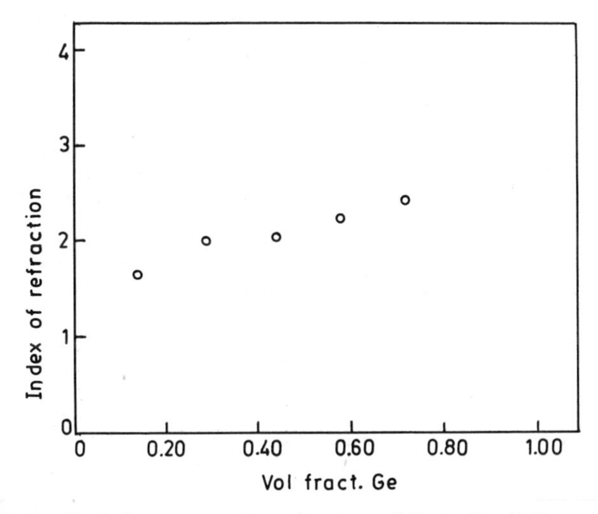

Fig. 5.10.16 Refractive index versus volume fractions of Ge in Ge–CaF_2 composite films. Adapted from Gittleman et al. (1977) (Ref. 362).

In both cases, a strong absorption band was observed at 12 μm which does not exist in the reflectance spectra of Si, Ge, or CaF_2. Table 5.10.2 lists solar absorptance and thermal emittance at different temperatures for Si and Ge composite films. Figure 5.10.19 shows the variation of conversion efficiency with solar concentration c at 300 and 500°C for Ge–CaF_2 and Si–CaF_2 films. Gittleman (361) reported that the dependence of performance

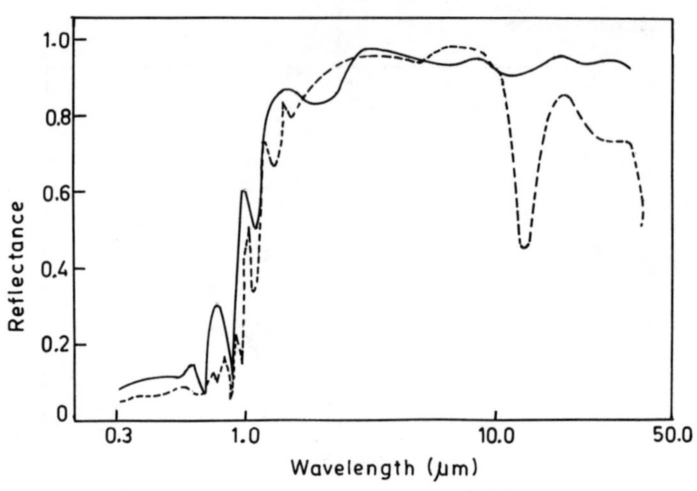

Fig. 5.10.17 Spectral reflectance for Ge–CaF_2 composite films having different thickness and 40 vol. % Ge and deposited on anodized Al on Si. Full line: 1.7 μm thick Ge–CaF_2 layer dashed line: 0.6 μm thick Ge–CaF_2 layer. Adapted from Gittleman et al. (1977) (Ref. 361).

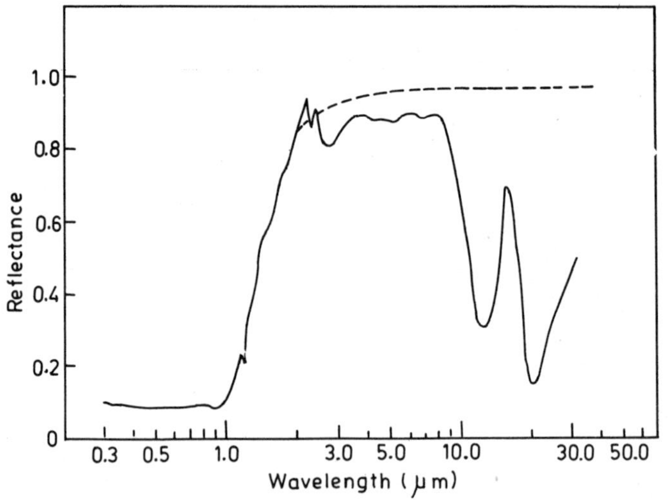

Fig. 5.10.18 Spectral reflectance of Si–CaF_2 5 μm thick composite film deposited on aluminum mirror. Adapted from Gittleman et al. (1977) (Ref. 361).

Table 5.10.2 Solar Absorptance and Thermal Emittance of Ge–CaF_2 Composite Films Having Different Thicknesses

Sample	α_s	ε_T(300°C)	ε_T(500°C)
1.7 μm Ge–CaF_2/Al	0.721	0.101	0.077
0.6 μm Ge–CaF_2/Al	0.648	0.061	0.053
Si–CaF_2/Al	0.7	0.1	—

Source: Adapted from Gittleman (1977) (Ref. 362).

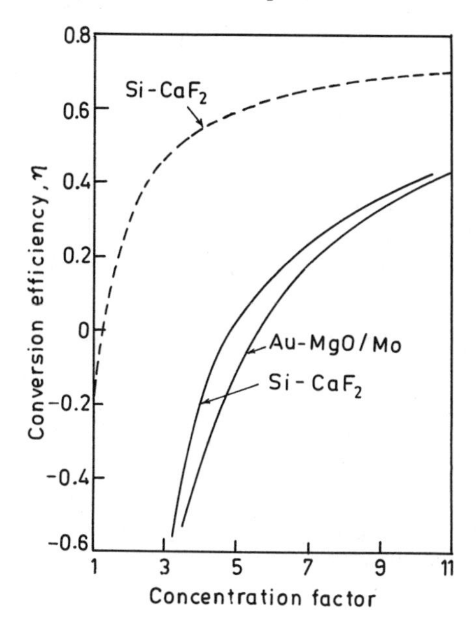

Fig. 5.10.19 Collection efficiency versus solar concentration for Si–CaF_2 films at 500°C and comparison with Au–MgO/Mo film. Adapted from Gittleman et al. (1977) (Ref. 361).

parameters on Si and Ge concentration is very weak and optimum performance was found to occur around 40 vol. % Ge in Ge–CaF_2 films.

5.11 Dielectric Constant of Composite Selective Surfaces

The dielectric constant of a composite material can be formulated on the basis of effective medium theory and Maxwell–Garnett theory. In the effective medium theory (347, 349) the two components, A and B of the composite material are treated in an equivalent manner. Grains of A and B are embedded in an effective medium whose dielectric constant is the same as that of composite material. The polarization of grains A and B, when embedded in the effective medium, average to zero, so that the theory is self-consistent.

In the Maxwell–Garnett theory it is assumed that the grains of one component are embedded in a matrix of the other component (i.e., grains of A in matrix B and grains of B in matrix A). Therefore, in contrast to effective medium theory, the two components A and B are not treated in an equivalent manner.

The Maxwell–Garnett theory takes into account the modification of the applied electric field at any point within the medium by dipole fields of the surrounding metal particles. The theory assumes that the metal particles are spherical and small compared to the wavelength of light but sufficiently large so that the macroscopic Maxwell equations can be applied. In the metal rich compositions, where metal particles coalesce, the Maxwell–Garnett theory breaks down. The Clausius–Mosotti equation for a collection of metal particles can be expressed as:

$$\frac{\varepsilon_{GM}(\omega) - 1}{\varepsilon_{GM}(\omega) + 2} = \frac{4}{3}\pi n \alpha(\omega) \tag{5.11.1}$$

where $\varepsilon_{GM}(\omega)$ is the frequency-dependent dielectric constant, n is the volume density, and $\alpha(\omega)$ is the polarizability. Equation 5.11.1 may be modified for the metal particles dispersed in a polarizable insulator of dielectric constant ε_i rather than in a vacuum. Thus in the calculations of local electric field, it is necessary to construct a spherical cavity filled with the insulator instead of completely evacuated cavity. The generalized Clausius–Mosotti equation may be expressed as:

$$\frac{\varepsilon_{GM}(\omega) - \varepsilon_i(\omega)}{\varepsilon_{GM}(\omega) + 2\varepsilon_i(\omega)} = \frac{4\pi n \alpha(\omega)}{3\varepsilon_i(\omega)} \tag{5.11.2}$$

The macroscopic dielectric theory is inapplicable for large metal particles. One can therefore substitute for $\alpha(\omega)$, the expression for the polarizability (364) of an isolated metal sphere immersed in a medium of dielectric constant $\varepsilon_i(\omega)$. The substitution gives

$$\frac{\varepsilon_{GM}(\omega) - \varepsilon_i(\omega)}{\varepsilon_{GM}(\omega) + 2\varepsilon_i(\omega)} = \frac{f[\varepsilon_m(\omega) - \varepsilon_i(\omega)]}{\varepsilon_m(\omega) + 2\varepsilon_i(\omega)} \tag{5.11.3}$$

where f is the volume fraction of metal and $\varepsilon_m(\omega)$ is the dielectric constant of metal. Cohen et al. (340) generalized the above equation for ellipsoidal shaped grains. The ellipsoids were assumed to be identical in shape and orientation with one of their principal axis parallel to the external field. The above expression may then be written as

$$\frac{\varepsilon_{GM}(\omega) - \varepsilon_i(\omega)}{L_m\varepsilon_{GM}(\omega) + (1 - L_m)\varepsilon_i(\omega)} = f\frac{\varepsilon_m(\omega) - \varepsilon_i(\omega)}{L_m\varepsilon_m(\omega) + (1 - L_m)\varepsilon_i(\omega)} \tag{5.11.4}$$

where L_m is the characteristic depolarization factor of the ellipsoids. ($L_m = \frac{1}{3}$ for spherical metal particles.)

The expression for $\varepsilon_{GM}(\omega)$ in the metallic region, where the structure of granular metal consists of dielectric grains dispersed in metal matrix, is obtained from eq. 5.11.4 by replacing $\varepsilon_m(\omega)$ by $\varepsilon_i(\omega)$, f by $1-f$, and L_m by L_i. The final expression may be given by

$$\frac{\varepsilon_{GM}(\omega) - \varepsilon_m(\omega)}{L_i\varepsilon_{GM}(\omega) - (1 - L_i)\varepsilon_m(\omega)} = (1-f)\frac{\varepsilon_i(\omega) - \varepsilon_m(\omega)}{L_i\varepsilon_i(\omega) - (1 - L_i)\varepsilon_m(\omega)} \tag{5.11.5}$$

where L_i is the effective depolarization factor of the insulating inclusions. The metal dielectric constant $\varepsilon_m(\omega)$ is modified to take account of the decrease in the conduction electron relaxation time τ due to the small particle microstructures (340). $\varepsilon_m(\omega)$ is decomposed into a free electron part $\varepsilon_m^f(\omega)$ and an interband part $\varepsilon_m^i(\omega)$.

$$\varepsilon_m(\omega) = \varepsilon_m^f(\omega) + \varepsilon_m^i(\omega)$$

where

$$\varepsilon_m^f(\omega) = 1 + i\frac{\bar{\omega}_p^2 \tau \omega^{-1}}{1 - i\omega\tau}$$

where $\bar{\omega}_p$ is the free electron plasma frequency and τ is conduction electron relaxation time. Cohen et al. (340) and Abeles (363) discussed several predictions, e.g., IR behavior, plasma frequency, and the presence of dielectric anomaly below the plasma frequency, of Maxwell–Garnett theory.

Self-consistent effective medium theory which is valid for all volume fractions f, was originally put forward by Bruggeman (351) and has since been republished by Odelevshii (365) and by Landauer (366). The effective medium permeability $\varepsilon_{\mathrm{BR}}$ is given by:

$$\varepsilon_{\mathrm{BR}} = \varepsilon_m(\omega)\frac{1 - f + \frac{1}{3}\sum_J f_J \alpha_J}{1 - f - \frac{2}{3}\sum_J f_J \alpha_J} \tag{5.11.6}$$

where subscript J denotes particles belonging to the J column in a size histogram centered at a diameter x_J and f_J's are a set of fractional filling factors normalized by $\sum_J f_J = f$. α_J is proportional to the polarizability of particles in the Jth class and for ellipsoidal particles is expressed by (367)

$$\alpha_J = \frac{1}{3}\sum_{i=1}^{3}\frac{\varepsilon_J - \bar{\varepsilon}}{\bar{\varepsilon} + L_i(\varepsilon_J - \bar{\varepsilon})} \tag{5.11.7}$$

L_i's are depolarization factors for the Jth particles and $\bar{\varepsilon}$ is the effective medium permeability. For a two-component mixture characterized by $\varepsilon_i(\omega)$ and $\varepsilon_m(\omega)$, the equation reduces to

$$f\frac{\varepsilon_i(\omega) - \varepsilon_{\mathrm{BR}}(\omega)}{\varepsilon_i(\omega) + 2\varepsilon_{\mathrm{BR}}(\omega)} + (1 - f)\frac{\varepsilon_m(\omega) - \varepsilon_{\mathrm{BR}}(\omega)}{\varepsilon_m(\omega) + 2\varepsilon_{\mathrm{BR}}(\omega)} = 0 \tag{5.11.8}$$

Recently Hunderi (368) gave another expression, for effective permeability, which holds good for small filling factors and may be written as

$$\varepsilon_{\mathrm{HU}}(\omega) = \varepsilon_m(\omega)\left(\frac{1 + \frac{1}{4}\sum_J f_J \alpha_J}{1 - \frac{1}{4}\sum_J f_J \alpha_J}\right)^2 \tag{5.11.9}$$

5.12 Quantum Size Effects (QSE)

Mancini and coworkers (369–373) reported that quantum size effects (QSE) occur in ultrathin films of degenerate semiconductors and result in high solar absorptance and simultaneous high thermal reflectance. The combination of a QSE material with a highly reflecting substrate can make a good selective absorber for photothermal conversion. In case of semiconductor materials, in order to achieve a good selective surface properties, a proper energy band gap is required. Truly speaking, the QSE furnish a process for varying with good degree of freedom the band gap of a thin film to optimize its banding to the solar spectrum.

In principal, QSE in solid crystalline films should occur when film thickness is comparable with the de Broglie electron wavelength. Under these conditions, the electron energy spectrum is characterized by the occurence of subbands separated by

$$\delta\varepsilon = \frac{\hbar^2\pi^2}{2m^*d^2}(2n - 1) \tag{5.12.1}$$

where m^* is the electron effective mass, d is the film thickness, and n is allowed to assume nonzero integer values. The formation of such subbands is shown in Fig. 5.12.1. Within each subband, the first allowed electron energy eigenvalue is thus shifted by $\hbar^2\pi^2/2m^*d^2$ with respect to the corresponding value in bulk sample. The critical length below which QSE have to be accounted

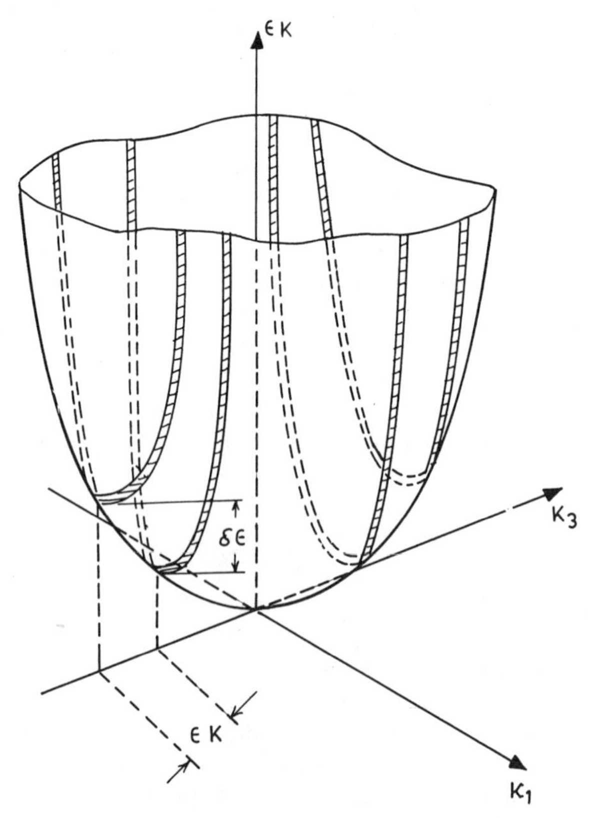

Fig. 5.12.1 Formation of subbands in a thin films in the presence of quantum size effect. Adapted from Giaquinta and Mancini (1976) (Ref. 373).

for is given by

$$d_c = \left(\frac{\pi}{2n} \right)^{1/3} \tag{5.12.2}$$

where n is the carrier concentration.

Generally most of the metals have high carrier density of the order of 10^{22}–10^{23} electrons cm^{-3} and therefore the critical length turns out to be of the order of tens of angstroms. The QSE has to be contrasted with semi-metals or degenerate semiconductors, where low carrier density ($\approx 10^{16}$ cm^{-3}) gives a critical length d_c of about 400–500 Å. Thus in the realm of QSE, the effective gap amplitude between the valence band and conduction

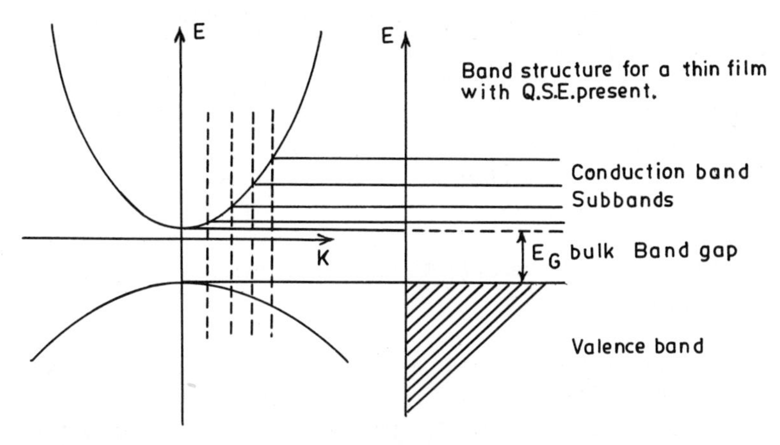

Fig. 5.12.2 The shift of a band gap in a degenerate semiconductor in the quantum size effect (QSE) regime. Adapted from Giaquinta and Mancini (1976) (Ref. 373).

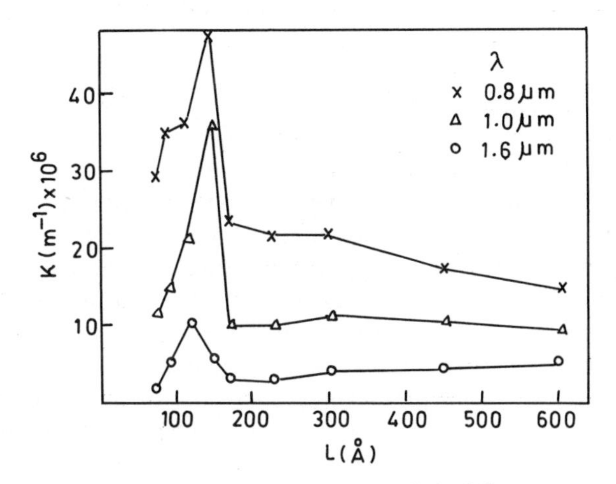

Fig. 5.12.3 Absorption coefficient of Al–InSb system versus InSb film thickness. Adapted from Giaquinta and Mancini (1976) (Ref. 373).

band in a degenerate semiconductor can be changed as a function of its thickness (Fig. 5.12.2). The QSE effect in vacuum deposited indium antimonide (InSb) on silver and aluminium substrates, has been extensively studied by the Catonia group (371). The (InSb) films were deposited in a vacuum by allowing the particles to fall onto the heated substrate. The absorption coefficient was measured for this film. Figure 5.12.3 shows results for various film thicknesses analyzed by Burrafato et al. (372) for various values of incident radiations. Similar effects may be investigated in semiconductors exhibiting sharp conduction band minima. This would require a low value of the ratio of electron effective mass to that of its free mass in the thin film. QSE may help in understanding the high solar energy absorption phenomenon in multilayer selective absorbers in which a ultrathin metallic film (40 Å in case of Mo) is sandwiched between two dielectric layers. The major disadvantage at present in exploiting QSE for solar selective surfaces is the continuity of composition, and the stability of the coating upon heating and cooling cycles and atmospheric exposure. To understand the exact mechanism and utilize the QSE, extensive studies are required for degenerate semiconductors and their films and multilayer stacks.

5.13 Selective Surfaces for Concentrating Systems by Magnetron Sputtering

For concentrating solar collector systems a selective surface having high temperature (400–500°C) stability is required for reasonably good efficiencies. An evaporated Al_2O_3–Mo–Al_2O_3 interference coating (solar absorptance = 0.85, thermal emittance = 0.11 at 500°C) with ≈ 400 Å thick layers has demonstrated excellent thermal stability up to 900°C (254). The sputtering process is appropriate for depositing thin multilayer coatings because of the precise control of thickness and uniformity.

Sputtering is an attractive process for the deposition of thin coatings because a large number of materials may be deposited by it. The sputtering process is a vacuum coating method wherein coating material is passed into the vapor phase by ion bombardment of a source consisting of the material in question. In this process the coating material is passed into vapor phase by a momentum exchange rather than a chemical or thermal process and therefore, virtually any material can be deposited. The basic problem in implementing the sputtering process is to maintain a plasma of sufficient intensity to provide a copious supply of ions over the surface of the target that has a geometry compatible with the desired substrate surface. For uniform coating thickness, the plasma must be of uniform density over the target surface and this intense plasma state must also be maintained in an argon gas that is at a low enough pressure to allow the sputtered atoms to pass through it freely without severe collisional scattering. Recent advances in sputtering technology (374) using magnetically confined plasmas in so-called magnetron geometry, where circulating sheet or ring currents are maintained over cylindrical or planar (374) targets, have significantly raised deposition rates and substrate areas. When an electron is in a magnetic field, its velocity component parallel to the magnetic field is unaffected. However its velocity component perpendicular to the magnetic field becomes a circular motion about the magnetic field line, if an electric field is applied perpendicular to magnetic field as shown in Fig. 5.13.1(a). The electron then moves in a spiral

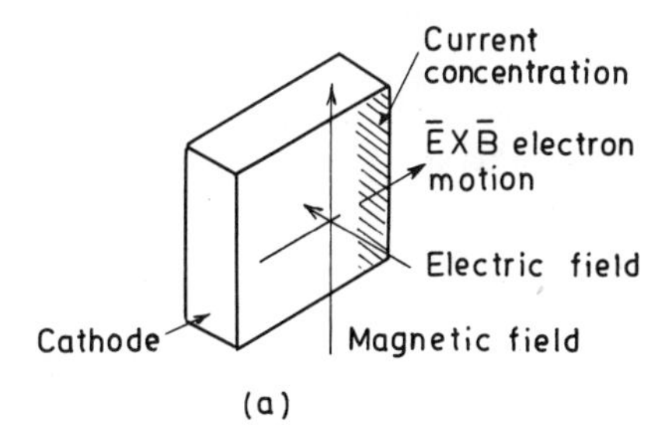

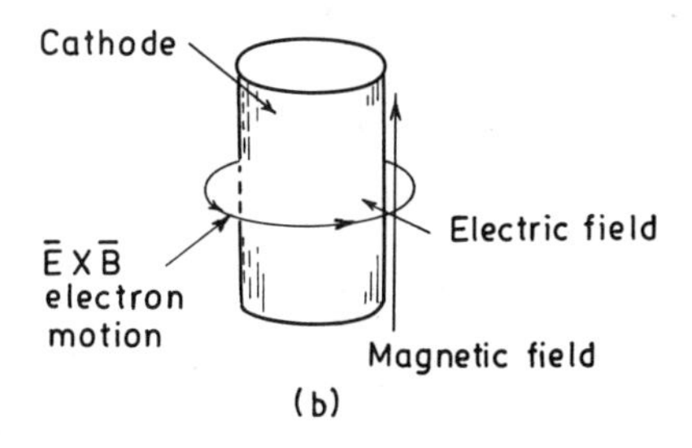

Fig. 5.13.1 Electromagnetic field interaction on planar and cylindrical cathodes. Adapted from Thronton (1974) (Ref. 374).

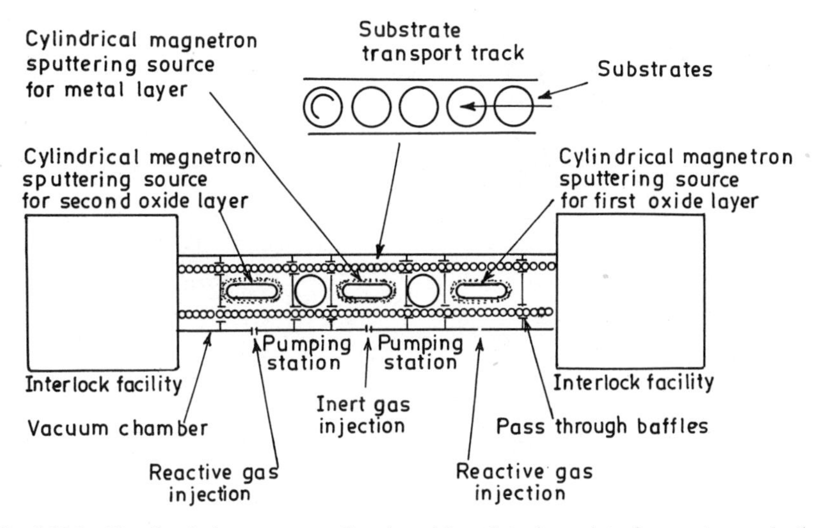

Fig. 5.13.2 Sketch of the apparatus for depositing three-layer interference type selective absorber coatings in tubular sections. Adapted from Thronton (1976) (Ref. 254).

perpendicular to both the electric and magnetic field directions as shown in Fig. 5.13.1(b).

Thornton (254) produced multilayer interference coatings consisting of 400 Å thick layers of an oxide, a metal, and an oxide. The coating was deposited on the outside of 3 ft long sections of 2 in, diameter tubing for use in a concentrating type solar collector. The setup is shown (254) in Fig. 5.13.2. The values of solar absorptance and thermal emittance were 0.85–0.9 and 0.2, respectively, for oxide–stainless–steel–oxide multilayer stacks.

5.14 Performance of Honeycomb Solar–Thermal Converters

Utilization of solar energy for heating and cooling of buildings requires the use of flat-plate collectors having reasonable efficiencies at temperatures

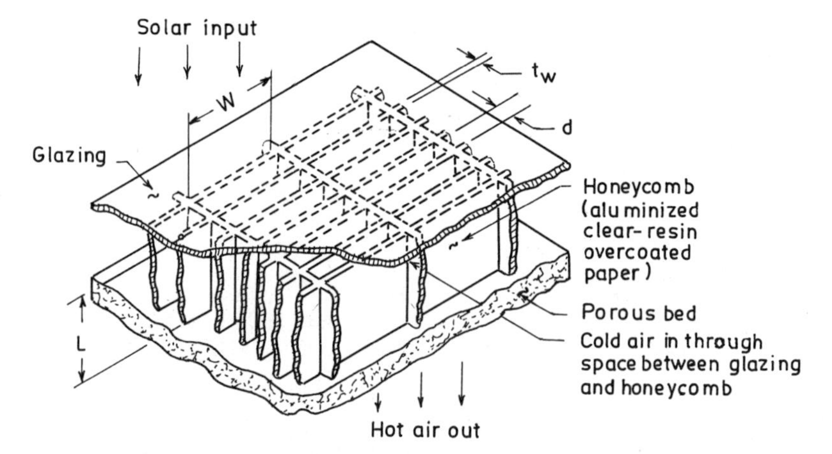

Fig. 5.14.1 Rectangular Honeycomb photothermal convert. Adapted from Buchberg et al. (1971) (Ref. 378).

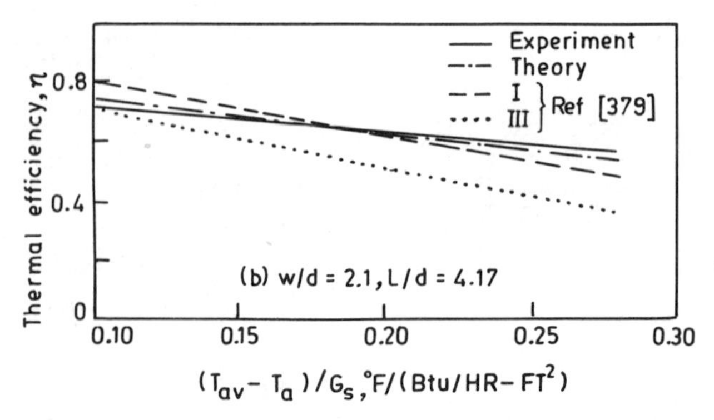

Fig. 5.14.2 Comparison of thermal efficiencies of experimental Honeycombs with theory and efficiencies of hearter I and III of Ref. 379. Adapted from Buchberg et al. (1971) (Ref. 378).

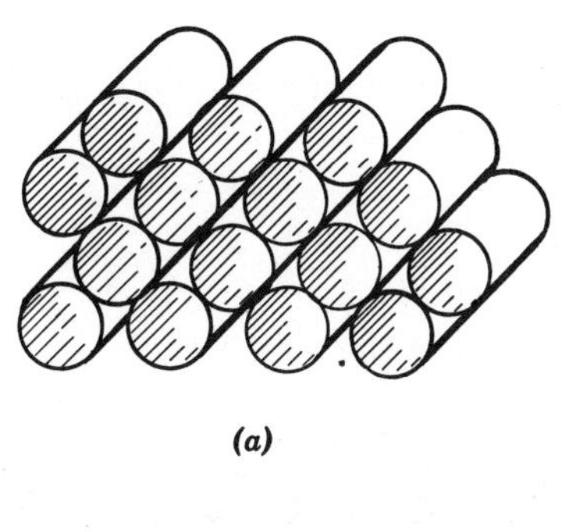

(a)

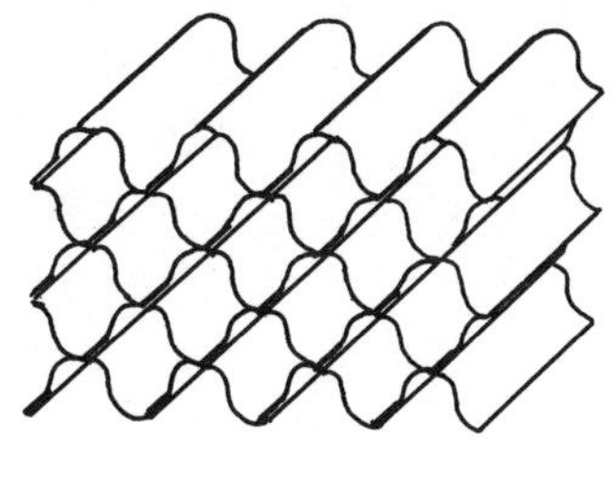

(b)

Fig. 5.14.3 Honeycomb structure based on (a) thin-walled tube; (b) corrugated thin glass sheets. Adapted from Mackenzie (1975) (Ref. 381).

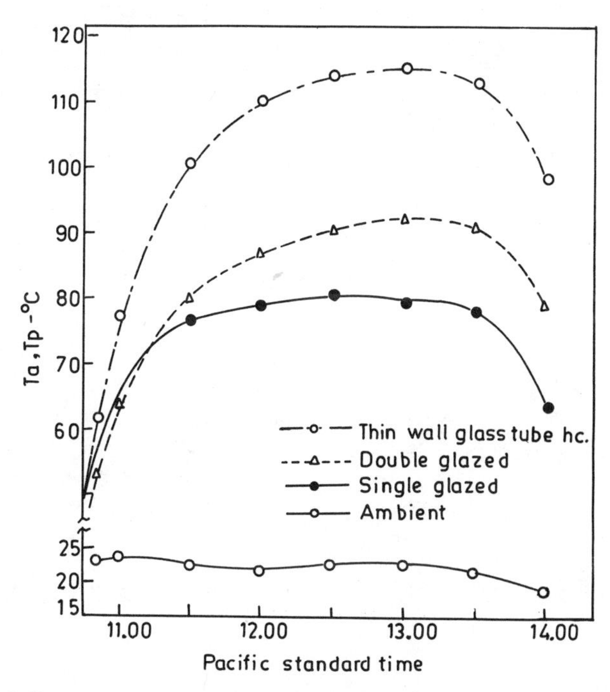

Fig. 5.14.4 Collector plate temperature for single and double glazed and thin wall glass tube Honeycomb in simple solar collector. Adapted from Mackenzie (1975) (Ref. 381).

Table 5.15.1 Solar Selective Properties of Various Types of Black Absorbing Selective Coatings

Serial No.	Material	Substrate	Fabrication Technique	Solar Absorptance α_S	Thermal Emittance, ε_T^a	Stability (°C)	Reference
1.	HfC	—	Bulk material	0.65	0.10		83
2.	NiS + ZnS (Black nickel)	Steel	Electrodeposited	0.88	0.10(100)	320	213
3.	NiS + ZnS	Nickel coated steel	Electrodeposited 2 layers	0.96	0.07(100)	280	220
4.	NiS + ZnS	Nickel Coated Steel	Electrodeposited 3 layers	0.91	0.14	—	382
5.	NiS + ZnS	Zinc	Electrodeposited	0.94	0.09(100)	200	215
6.	NiS + ZnS	Stainless steel	Electrodeposited	0.84	0.18	230	210
7.	CrO_x	Nickel plated steel	Electrodeposited	0.96	0.11	300	217–221
8.	CrO_x	Aluminium foil	Electrodeposited	0.964	0.023	—	332
9.	CrO_xCu	—	Paint coating	0.92	0.30(100)	—	
10.	$Cr_2O_3 + Cr$	—	RF sputtering	0.92	0.08(100)	400	293
11.	CuO	Al Sheet	Spray pyrolysis	0.93	0.11	—	235
12.	CuO	Al sheet	Electrodeposited				
			3 min	0.79	0.05	—	207
			8 min	0.89	0.17	—	207
13.	CuO	Al Sheet	Electrodeposited	0.90	0.15	—	211
14.	CuO/Au	—	Paint	0.80	0.06	200	382
15.	$CuO/Ag/Rh_2O_3$	—	Paint	0.90	0.10	400	383
16.	CuO + Fe + Mn	—	Paint	0.92	0.13(100)	150	382
17.	CuO	Copper	Electrodeposition	0.91	0.05	210	210

18.	Fe_xO_y	Steel	Electroless Deposition	0.85	0.10	—	236
19.	Fe_xO_y	Steel	Heating steel in air	0.88	0.12	—	211
20.	Fe_xO_y	Steel	DC Reactive sputtering	0.80	0.2(150)	350	288
21.	CoO_x	—	Anode Coelectrode-position	0.93	0.24(260)	—	85
22.	CoO_x	Nickel	Electrodeposition and heating	0.87	0.07(60)	—	251
23.	WO_3	Nickel	Rf Reactive sputtering	0.83	0.07	400	251
24.	$WO_3 + Al_2O_3$	Nickel	Rf Reactive sputtering	0.93	0.09	400	251
25.	Cu_2S	Copper	Chemical conversion	0.79	0.2(200)	—	242
26.	Converted zinc	Zinc	Chromate conversion	0.79	0.07	—	255
27.	Converted zinc	Zinc	Chloride conversion	0.93	0.08	—	255
28.	ZnO	Steel coated with bright zinc	Chemical brightening and etching	0.95	0.08	100	255
29.	Colored stainless steel	Stainless steel	Chemical conversion	0.93–0.62	0.10	—	258
30.	Alcoa 655	Al sheet	Chemical conversion	0.93	0.35	—	266
31.	Si	Ag	CVD	0.80	0.05(100)	500	276
32.	Si + 0.5 μm Ge	Ag	CVD	0.892	0.0411	—	277
33.	PbS	Al sheet	Vacuum evaporation	0.94	—	—	279
34.	PbS	Al painted Al sheet	Spray pyrolysis	0.93	0.15	200	282
35.	PbS	Al evaporated Al sheet	Vacuum evaporation	0.97	0.10	300	282
36.	Chromium carbide	Bulk copper	RF Reactive sputtering	0.80	0.02	500	284
37.	Chromium carbide	Bulk copper and sputtered metal	RF Reactive sputtering	0.85	0.05	500	284

Table 5.15.1 (*Continued*)

Serial No.	Material	Substrate	Fabrication Technique	Solar Absorptance α_S	Thermal Emittance, ε_T^a	Stability (°C)	Reference
38.	Chromium silicide	Bulk copper	RF Reactive sputtering	0.765	0.02	600	284
39.	PbS	Aluminium	Paint coating	0.90	0.37	300	278
40.	PbS	Aluminium	Paint coating	0.96	0.75	300	227
41.	Si	Al	Paint coating	0.83	0.70	300	227
42.	Ge	Al	Paint coating	0.91	0.79	300	227
43.	Inorganic pigment	Al	Paint coating	0.92	0.36	100	211
44.	Organic pigment	Al	Paint coating	0.62	0.12	100	221
45.	CuO on Zn	Al	Paint coating	0.95	0.42	250	295
46.	CuS + PbS on Zn	Al	Paint coating	0.96	0.49	250	295
47.	CuO on Zn	Al	Paint coating	0.94	—	—	294
48.	MgF_2–Mo–CeO_2	Mo sheet	Vacuum evaporation	0.85	0.053	400	298
49.	AMAMAMA	Molybdenum	Vacuum evaporation	0.91	0.085(260)	400	299
50.	SiO_2–Mo–SiO_2	Molybdenum	Vacuum evaporation	0.86	0.08		306
51.	Nickel dendrites	Stainless-steel tungsten	CVD	0.99	0.26(550)	600	319
52.	Nickel dendrites	Sapphire stainless-steel tungsten	CVD	0.95	0.6(100)	300	318
53.	Gold black	Copper	Gas evaporation	0.88	0.10		324
54.	Au–MgO	Copper	Cosputtering	0.93	0.04		343
55.	Au–MgO	Stainless steel	Cosputtering	0.93	0.10		343

56.	$Cr-Cr_2O_3$	Copper	Sputtering	0.92	0.08	—	293
57.	$Ni-Al_2O_3$	Copper	Coevaporation	0.91	0.10	—	360
58.	$Pt-Al_2O_3$	Copper	Coevaporation	0.94	0.19	—	360
59.	$Ge-CaF_2$	Aluminium	Cosputtering	0.721	0.101	—	362
60.	$Si-CaF_2$	Aluminium	Cosputtering	0.7	0.1	—	361
61.	Al/Ge/SiO	—	Vacuum evaporation	0.79	0.012	—	384
62.	Al/PbS/SiO	—	Vacuum evaporation	0.89	0.018	—	384
63.	Ni/Ge/SiO	—	Vacuum evaporation	0.88	0.035	240	384
64.	Ni/PbS/SiO	—	Vacuum evaporation	0.93	0.043	—	384
65.	Cr/Ge/SiO	—	Vacuum evaporation	0.93	0.11	240	384
66.	Cr/PbS/SiO	—	Vacuum evaporation	0.94	0.12	240	384

[a] The figures in the parentheses under ε_T denote the temperatures (in degrees centigrade) at which the emittance has been measured.

in the vicinity of 100–110°C. Ordinary solar flat-plate collectors have relatively low efficiencies at these temperatures due to excessive convection and reradiation losses. A honeycomb structure placed between the absorber plate and the transparent cover plate of a flat-plate collector suppresses the convection and radiation losses. This permits the absorber plate to attain high temperatures. In 1929 Veinberg (375) experimented with cell structures formed by stacked wire meshes or vertical sheets of lacquered, corrugated paper laid over the absorbing plate. Francia (376) demonstrated the effectiveness of cell structures laid over medium- and low-temperature solar absorbers. Following Francia's investigations, Hollands (377) presented a simplified analysis of honeycomb solar collectors. Further, Buchberg et al. (378) have done extensive studies of radiation transfer and free convection on the performance of rectangular honeycomb solar thermal converters. Theoretical expressions are presented for the performance of both nonflow and air transpiration converters. The analysis was validated by experimental results obtained with a 1 ft^2 test module, assembled in a glazed and well-insulated housing mounted on a tilting platform for the desired orientation. The honeycomb structure investigated, shown in Fig. 5.14.1, consists of an arrangement of multiple rectangular cells with highly specularly reflecting walls coated with a dielectric film. The film is transparent to solar radiation and absorbs long wavelength radiations. The measurements were made with three configurations (1) $L/d = 7.11$, $W/d = 3.4$ for $L = 1.5$ in.; (2) $L/d = 4.67$, $W/d = 3.4$ for $L = 1.0$ in.; and (3) $L/d = 4.17$, $W/d = 2.1$ with $L = 1.5$ in. ($L, d,$ and w are defined in Fig. 5.14.1). Figure 5.14.2 compares the mean theoretical predictions with the experimental mean curves. The figure also shows the comparison with the data given in Ref. 379 for the corrugated air heaters I and III. The agreement between the theory and experiment is excellent and it is obvious that the honeycomb air heaters are more efficient than the corrugated air heater at higher absorber temperatures. Cunnington and Stread (380) at Lockhead Missiles and Space Co. have demonstrated a significant improvement of efficiencies with the thin-walled transparent honeycomb structures based on organic plastics mylar. But later on Mackenzie (381) investigated transparent thin-walled honeycomb based on inorganic glasses because of the many undesirable characteristics of organic plastics such as deterioration in the UV, low glass transition temperature, and absorption of water. Figure 5.14.3(a, b) shows a thin-walled structure based on glass tubings and corrugated sheets proposed by Mackenzie. Figure 5.14.4 compares the thermal performance of simple collector with thin-walled glass tube honeycomb, single and double glazed. For a nonselective black absorber, such honeycombs are much more efficient than a glazed flat-plate collector. Although a honeycomb structure is useful in suppressing thermal IR radiation from the collector surface, a function similar to a solar selective surface, their use is not cost effective.

6

Conclusion and Recommendations

6.1 Transparent Conducting Coatings

Transparent heat mirror coatings that transmit in the visible and reflect in the IR have important applications in the collection and conversion of solar energy. Unlike the absorber coatings, which are applied on the surface of the collector, the transparent heat mirror coating is applied to the cover of flat-plate collector. The heat mirror permits the transmission of solar radiation but reflects the IR. Drude mirrors, i.e., idealized conductors whose optical properties depend on the free electron contributions, are of interest because they have a high transparency at wavelengths below the plasma edge with high reflectivity at longer wavelengths. The position and sharpness of the edge are determined by the mobility and concentration of the charge carriers. Doped tin oxide, indium oxide, indium tin oxide, and cadmium stannate offer good possibilities for transparent heat mirror applications in solar energy collectors. The coatings have good thermal stability up to 300°C, have excellent adhesion to glass, have high resistance to chemical attack, and are easily incorporated in the collector by replacing the cover. An ordinary collector can be converted to an advanced type simply by replacing the cover. The feasibility of large-area coatings has been established. The indium tin oxide and cadmium stannate are the two promising materials and further development work on these and effort to find similar new materials should continue.

There has been considerable interest in the past few years in the use of transparent conducting layers as constitutive elements in photovoltaic devices. Such devices are of current interest because they represent a potentially low cost method for fabricating large-scale solar energy conversion arrays. The cost reduction of such cells is in the junction formation step and also in eliminating the antireflection layer. ITO/Si devices of 13% efficiency have been reported. Due to low-temperature processing they open up the possibility of using polycrystalline silicon in future technology. The future work should aim at achieving 10% efficiencies with oxide–

semiconductor–insulator–polycrystalline silicon solar cells. Techniques for growing stoichiometric ultrathin oxides have to be developed to achieve highest conversion efficiencies.

6.2 Plated Coatings

Black nickel, a nickel–zinc sulphide complex and black chrome, a mixture of chrome and chromium oxides are the most promising electroplated absorber coatings for low-temperature (150 to 300°C) solar collector applications. The best values for α and ε obtained to-date for black nickel on nickel plated substrates are 0.95 and 0.07, respectively. These coatings were found to be stable to temperatures of 300°C in a vacuum. The coatings were also stable under UV irradiation in a vacuum, but were unstable when irradiated in air. The coatings however, failed under high temperature and humidity.

Black chrome has good optical properties ($\alpha = 0.95$ and $\varepsilon = 0.09$) and is the most durable. With a 0.5 mil undercoat of nickel, the coating life is estimated to be more than 20 years. The nickel layer, however, adds considerably to the cost (the cost of nickel layer is about half of the total cost of the coating) and further work towards finding a cheap substitute for the nickel layer are needed so as to make the black chrome cost effective. Long-term outdoor testing of these coatings also needs to be done. Iron oxide on iron has the lowest cost among the electroplated absorbers. Optical properties in the range of $\alpha = 0.85$ and $\varepsilon = 0.1$ are obtained. The durability of these coatings is considerably enhanced with organic overcoats and this material is therefore a good candidate for low-temperature applications.

6.3 Paint Coatings

Paint coatings have recently received considerable attention as low-cost large-area coatings for low-temperature applications in solar energy collectors. The selective paints consist of particulates of semiconductors in a suitable thermally stable and IR transparent binder. The spectral property of a selective paint depends upon the optical properties of pigments, the particles size, and the multiple scattering effects within the pigment–binder composite. Among the pigments studied, the iron–manganese–copper oxide has the best optical properties. Selective coatings made with this metal oxide have $\alpha = 0.92$ and $\varepsilon = 0.13$ when in silicone binders and $\alpha = 0.92$ and $\varepsilon = 0.10$ when an aliphatic urethane binder is used. The selective optical properties of solar paints are very sensitive to pigment solar absorptance, coating thickness, and the pigment-to-volume concentration. Further work needs to be done to optimize these parameters. Due to the IR absorption of the binder materials, the IR emittance of paint coatings is high. More work needs to be done on pigments and binders to improve the optical properties. The organic binders of selective paints improve the environmental stability but the mechanical durability of the paint coatings is very poor. More work is needed to improve the adhesion of paint coatings. At present little is known about weathering resistance, humidity, and thermal cycling resistance of paint coatings. Long-term outdoor tests are needed to know the potential of paint coatings. Suitable large-scale production processes have to be developed. Improvements in the optical properties and long-term durability tests are needed to establish the potential of metal dust pigmented solar selective paint coatings which appear to be a promising candidate among the paint coatings.

6.4 High-Temperature Selective Surfaces

Composite material films consisting of ultrafine metal particles in an insulating matrix have received considerable attention during the past few years as high-temperature selective surfaces. X-ray diffraction investigations on black chrome selective black absorbers suggested that the material consists of polycrystalline Cr_2O_3 and amorphous Cr. The optical properties are affected by the crystallization of amorphous chromium at elevated temperatures. The degradation is also caused by the oxidation of chromium at 350°C in air. MgO/Au cermet films are stable up to 400°C. Structural studies revealed that both MgO and Au are crystallites smaller than about 200 Å. The cermet is essentially transparent in the IR and low emissivity is due to the metal base. The best optical properties obtained are $\alpha = 0.9$ and $\varepsilon = 0.1$. The Ni/Al_2O_3 composite films produced by simultaneous evaporation of the metal and dielectric by an electron-beam evaporator were stable up to 500°C. Greater efficiency can be obtained in graded composite films. The best optical properties obtained in graded Ni/Al_2O_3 gave $\alpha = 0.94$ and $\varepsilon = 0.1$. Adequate test data on cermets is not yet available and further tests under load conditions need to be done.

If in composite films, the particles are assumed small compared to the wavelength of radiation, and the particles are far enough apart to ensure independent scattering, the Lorentz local field correction is applicable and the bulk optical constants can be used, and the theory of Maxwell and Garnett can be used to interpret the experimental data. While the optical properties of Ni/Al_2O_3 and W/MgO cermet films agreed fairly well with Maxwell–Garnett theory, the optical properties of Au/MgO were in disagreement with the theory. The deviation is possibly caused by the effects of particle interactions and multiple scattering, size and shape of the particles, and volume fractions of cermet components.

6.5 Magnetron Sputtering

Magnetron sputtering is an attractive process for producing selective surfaces. In this technique, a magnetic field is applied parallel to the surface of the cathode. The magnetic field increases the ionization for a given gas pressure and allows the discharge to be run at lower gas pressures, resulting in an increased rate and efficiency of coating. Since the coating material is passed into the vapor phase by a momentum exchange, rather than a chemical or thermal process, virtually any material is a candidate for coating. In addition, the sputter discharge is far more stable and controllable. A three-layer oxide–metal–oxide coating can be deposited on tubular sections for use in concentrated collectors. Sputtering is particularly attractive for such applications, since the large-area capability permits relatively high volume production at moderate deposition rates. However, further work on this technique is needed to reduce the cost of large-area coatings.

References

1. J. C. C. Fan, F. J. Bachner, G. H. Foley, and P. M. Zavracky, *Appl. Phys. Lett.*, **25**, 693 (1974).
2. J. C. C. Fan and F. J. Bachner, *Appl. Optics*, **15**, 1012 (1976).
3. J. C. C. Fan, T. B. Read, and J. B. Goodenough, in *Proc. 9th Intersociety Energy Conversion Eng. Conf., San Francisco, Cal.*, August 26–30, 1974.
4. J. C. C. Fan, *Adv. Chem.*, **163**, 149 (1977).
5. S. Yoshida, *Appl. Optics*, **17**, 145 (1978).
6. S. Yoshida, *Oyo Butsuri*, **45**, 477 (1976).
7. J. H. Apfel, *J. Vac. Sci. Technol.*, **12**, 1016 (1975).
8. R. M. Winegarner, in *Proc. ISES and SES Canada*, Winnipeg, Canada, August 1976, K. W. Boer, Ed., Vol. 6, p. 339.

8a. R. M. Winegarner, *Opt. Spectra*, 12 (June 1975).

9. G. Haacke, *Appl. Phys. Lett.*, **28**, 622 (1976).
10. G. Haacke, *Appl. Phys. Lett.*, **30**, 380 (1977).
11. G. Haacke, *J. Appl. Phys.*, **47**, 4086 (1976).
12. G. Haacke, H. Ando, and W. E. Mealmaker, J. *Electrochem. Soc.*, **124**, 1923 (1977).
13. R. S. Sennett and G. D. Scott, *J. Opt. Soc. Am.*, **40**, 203 (1950).
14. F. W. Reynolds and G. R. Stilwel, *Phys. Rev.*, **88**, 418 (1952).
15. Corning Glass Works, "The Pyrex Brand Heat Shield," PE 23.
16. T. B. Read, "Solid State Research Report," No. 1969/1, Lincoln Laboratory, MIT, p. 21 (1969).
17. R. Groth and E. Kauer, *Phil. Tech. Rev.*, **26**, 105 (1965).
18. M. V. Schneider, *Bell Syst. Tech. J.*, **45**, 1611 (1966).
19. J. Holland and G. Siddal, *Brit. J. Appl. Phys.*, **9**, 359 (1958).
20. E. J. Gillham and J. S. Preston, *Proc. Phys. Soc. London Sect. B*, **65**, 649 (1952).
21. E. J. Gillham, J. S. Preston and B. E. Williams, *Phil. Mag.*, **46**, 105 (1955).
22. M. Garbuny, *Optical Physics* (Academic, New York, NY 1965), p. 36.
23. J. L. Vossen, *Phys. Thin Films*, **9**, (1977).
24. G. Haacke, *Ann. Rev. Mat. Sci.*, **7**, 73 (1977).
25. L. Holland, *Vaccum Deposition of Thin Films* (Wiley, New York, NY, 1958), pp. 492–509.
26. Z. M. Jarzebski and J. P. Marton, *J. Electrochem Soc.*, **123**, 199C-299C-333C (1976).
27. R. W. G. Wyckoff, *Crystal Structures* (Interscience, New York, NY, 1951).
28. W. H. Baner, *Acta Cryst.*, **9**, 515 (1956).
29. L. Pauling, *The Nature of Chemical Bond* (Cornell University Press, New York, NY, 1960).

30. S. Samson and C. G. Fonstad, *J. Appl. Phys.*, **44**, 4618 (1973).

31. R. E. Aitchison, *Austral. J. Appl. Sci.*, **5**, 10 (1954).

32. J. A. Aloaf, V. C. Marcotte, and N. J. Chou, *J. Electrochem. Soc.*, **120**, 701 (1973).

33. R. Summitt, J. A. Marley, and N. F. Borrelli, *J. Phys. Chem. Sol.*, **25**, 1465 (1964).

34. A. Fischer, *Z. Naturforsch.*, **A9**, 508 (1954).

35. A. R. Peaker and B. Horsley, *Rev. Sci. Instrum*, **42**, 1825 (1971).

36. A. Rohatgi, T. R. Vinerito, and L. H. Slack, *J. Am. Ceramic Soc.* **57**, 278 (1974).

37. T. Arai, *J. Phys. Soc. Japan*, **15**, 916 (1960).

38. I. Imai, *J. Phys. Soc. Japan*, **15**, 937 (1960).

39. G. Redaelli, *Appl. Optics*, **15**, 1122 (1976).

40. G. Murken and M. Tromel. *Z. Anorg. Allg. Chem.*, **397**, 117 (1973).

41. J. Kane, H. P. Schweizer, and W. Kern, *J. Electrochem. Soc.*, **123**, 270 (1976).

42. J. F. Jordan, 2nd *Natl. Sci. Found. Photovoltaic Conv. Res. Prog. Rev.*, Philadelphia, PA, 1974, p. 214.

43. R. L. Weiher and R. P. Ley, *J. Appl. Phys.* **37**, 299 (1966).

44. H. K. Muller, *Phys. Statics Sol.* **27**, 723 (1969).

45. Ahmar Raza, O. P. Agnihotri, and B. K. Gupta, *J. Phys. D: Appl. Phys.*, **10**, 1871 (1977).

46. R. Groth, *Phys. Stat. Sol.*, **14**, 69 (1966).

47. H. Kostlin, R. Jost, and W. Lems, *Phys. Stat. Sol. (a)*, **29**, 87 (1975).

48. H. J. J. Van Boort and R. Groth, *Phil. Tech. Rev.*, **29**, 17 (1968).

49. O. P. Agnihotri, A. K. Sharma, B. K. Gupta, and R. Thangaraj, *J. Phys. D: Appl. Phys.*, **11**, 643 (1978).

50. J. C. C. Fan and F. J. Bachner, *J. Electrochem. Soc.*, **122**, 1719 (1975).

51. J. L. Vossen, RCA Rev., **32**, 289 (1971).

52. D. B. Fraser and H. D. Cook, *J. Electrochem. Soc.*, **119**, 1368 (1972).

53. J. M. Pankratz, *J. Electron. Mat.*, **1**, 182 (1972).

54. W. W. Molzen, *J. Vac. Sci. Technol.*, **12**, 99 (1975).

55. E. Burstein, *Phys. Rev.*, **93**, 632 (1954).

56. A. J. Smith, *Acta Cryst.*, **13**, 749 (1960).

57. M. Tromel, *Z. Anorg. Allg. Chem.*, **371**, 237 (1969).

58. A. J. Nozik, *Phys. Rev.*, **86**, 453 (1972).

59. G. Haacke, W. E. Mealmaker and L. A. Siegel, *Thin Solid Films*, **55**, 67 (1978).

60. P. Lloyd, *Thin Solid Films*, **41**, 113 (1977).

61. O. P. Agnihotri, B. K. Gupta, and A. K. Sharma, *J. Appl. Phys.*, **49**, 4540 (1978).

62. H. Y. Fan, *Rep. Prog. Phys.*, **19**, 122 (1956).

63. G. E. McDonald, NASA, TMX-3136, NASA-Lewis Research Center (1974).

64. H. Tabor, in *Trans. Conf. Use of Solar Energy*, Tucson, AZ, October 31, 1955, E. F. Carpenter, Ed. (Univ. of Arizona Press, Tucson, AZ, 1955), Vol. 2, Part 1, Sec. A, p. 1.

64a. J. T. Gier and R. V. Dunkle, in *Trans. Conf. Use of Solar Energy*, Tucson, AZ, October 31, 1955, E. F. Carpenter, Ed. (Univ. of Arizona Press, Tucson, AZ, (1955) Vol. 2, Part I, Sec. A, p. 41.

65. L. Harris, R. T. Mcginnies, and B. M. Siegel, *J. Opt. Soc. Am.*, **38**, 582 (1948).

66. A. H. Pfund, *Rev. Sci. Instrum.*, **1**, 397 (1930).

67. A. H. Pfund, *J. Opt. Soc. A.M.* **37**, 558 (1947).

68. C. M. Horwitz, *Optics Commun.*, **11**, 210 (1974).

69. S. Ramo, J. Whinnery, and T. Van Duzer, *Fields and Waves in Communication Electronics* (Wiley, New York, NY, 1965).
70. J. C. C. Fan, F. J. Bachner, and R. A. Murphy, *Appl. Phys. Lett.*, **28**, 440 (1976).
71. M. I. Kontorovich, V. Yu, Petrunkin, N. A. Yesephkina, and M. I. Astrakhan, *Radio Eng. Electro. Phys.*, **7**, 223 (1962).
72. Gier Dunkle Instruments, Inc., Santa Monica, CA 90404.
73. J. A. Duffie and W. A. Beckman, *Solar Energy Thermal Processes* (Wiley, New York, NY, 1974).
74. R. Siegel and J. R. Howell, *Thermal Radiation Heat Transfer* (McGraw-Hill, New York, NY, 1972).
75. International Commission of Illumination, Publication CIE No. 20, TC-2.2 (CIE, Paris, 1972).
76. R. Campana and J. Rose, "Power Units for Spacecraft" (Russian Translation), in *Mir*, Moscow, USSR, 1964, p. 131.
77. A. V. Sheklein, *Geliotekhnika*, No. 1, 42 (1968).
78. B. P. Kryzhanovskii, *Opt. delil Mekh. Prom.*, No. 10 (1960).
79. A. V. Sheklein, N. B. Rekhant, E. A. Ihukovskaya, S. V. Yurkova, and M. A. Banlina, *Geliotekhnika*, No. 4, 57 (1966).
80. M. A. Alikhodzhaeva, L. V. Gudkov, and A. V. Skeklein, *Geliotekhnika*, No. 3, 19 (1965).
81. T. F. Irvine, J. P. Hartnett, and E. R. G. Eckert, *Solar Energy*, **2**, Nos. 3–4, 12 (1958).
82. R. B. Gillette, *Solar Energy*, **4**, No. 4, 24(1960).
83. B. O. Seraphin and A. B. Meinel in, *Optical Properties of Solids: New Developments* B. O. Seraphin, Ed. (North Holland Publishing Co., Amsterdam, 1976), Chap. 17.
84. G. E. McDonald, *Solar Energy*, **17**, 119 (1975).
85. C. M. Lampert, *Solar Energy Mat.*, **1**, (Jan. 1979).
86. L. N. Vladimirova, B. A. Garf, and A. V. Sheklein, in *Semiconductor Solar Energy Converters*, V. A. Baum, Ed. (Consultants Bureau, New York, NY, 1969), p. 209.
87. V. P. Isachenko, V. A. Osipova, and A. S. Sukomel, *Heat Transfer* (in Russian) (Energiya, Moscow, USSR, 1965).
88. M. A. Mikheev, *Fundamentals of Heat Transfer* (in Russian) (Gosenergoizdat, Moscow, USSR, 1956).
89. R. D. Goodman and A. G. Menke, *Solar Energy*, **17**, 207 (1975).
90. S. S. Nandwani, S. C. Mullik, P. K. Gogna, K. L. Chapra, and C. L. Gupta, Technical Report TR-78-S-6, Center of Energy Studies, Indian Institute of Technology, New Delhi.
91. J. B. DuBow, D. E. Burk, and J. R. Sites, *Appl. Phys. Lett.*, **29**, 494, 631 (1976).
91a. J. B. DuBow, J. Shewchun, C. Wilmsen, and W. S. Duff, in *Proc. ISES*, New Delhi, India, January 1978, F. de Winter and M. Cox, Eds. (Pergamon Press, New York, NY, 1978), p. 631.
92. E. Y. Wang and R. N. Legge, *IEEE Trans. Elect. Devices* **ED-25**, 800 (1978).
93. R. H. Bube, *Proc. Soc. Photo-optical Inst. Eng.*, **114**, 7 (1977).
94. T. Feng, C. Fishman, and A. K. Ghosh, in *13th IEEE Photovoltaic Spec. Conf.*, New York, NY, 1978, p. 519.
95. J. Shewchun, R. Singh, D. Burk, M. Spitzer, J. Loferski, and J. DuBow, in *13th IEEE Photovoltaic Spec. Conf.*, New York, NY, 1978, p. 528.
96. R. J. Stirn and Y. C. M. Yeh, *Appl. Phys. Lett.*, **27**, 95 (1975).
97. A. K. Ghosh, C. Fishman, and T. Feng, in *Proc. ISES*, New Delhi, India, January 1978, F. de Winter and M. Cox, Eds. (Pergamon Press, New York, NY, 1978).

98. S. J. Fonash, *J. Appl. Phys.*, **46**, 1286 (1975).

99. J. Shewchun, R. Singh, and M. A. Green, *J. Appl. Phys.*, **48**, 765 (1977).

99a. D. L. Pulfrey, *IEEE Trans. Elect. Devices*, **ED-25**, 1308 (1978).

100. R. Singh and J. Shewchun, *J. Vac. Sci. Technol.*, **14**, 89 (1977).

101. J. Shewchun, J. DuBow, A. Myszkowski, and R. Singh, *J. Appl. Phys.*, **49**, 855 (1978).

102. R. Singh and J. Shewchun, *Appl. Phys. Lett.*, **28**, 512 (1976).

103. T. Mizrah and D. Adler, *Appl. Phys. Lett.*, **29**, 682 (1976).

104. T. Mizrah and D. Adler, *IEEE Trans. Elect. Devices*, **ED-24,** 458 (1977)

105. E. Y. Wang and R. N. Legge, in *Proc. 12th Photovoltaic Spec. Conf.*, New York, NY, 1976, p. 967.

106. R. L. Anderson, *Appl. Phys. Lett.*, **27**, 691 (1975).

107. J. C. Manifacier and L. Szepessy, *Appl. Phys. Lett.*, **31**, 459 (1977).

107a. M. Perotin, L. Szepessy, J. C. Manifacier, P. Parto, J. P. Fillard, and M. Savelli, in *Coll. Intl. Elec. Solarire*, Toulouse, France, March 1–5, 1976, p. 481.

108. S. W. Lai, S. L. Franz, G. Kent, R. L. Anderson, J. K. Clifton, and J. V. Masi, in *11th IEEE Photovoltaic Spec. Conf.*, Scottsdale, AZ, 1975, p. 398.

109. J. P. Schunck and A. Coche, *Appl. Phys. Lett.*, **35**, 863 (1979).

110. K. Kajiyama and Y. Furukawa, *Jap. J. Appl. Phys.*, **6**, 905 (1967).

111. T. N. Nishinio and Y. Hamakawa, *Jap. J. Appl. Phys.*, **9**, 1085 (1970).

112. K. J. Bachmann, W. R. Sinclair, F. A. Thiel, H. Schreiber Jr., P. H. Schmidt, E. G. Spencer, W. L. Feldman, E. Buehler, and K. S. Sree Harsha, in *13th IEEE Photovoltaic Spec. Comf.*, New York, NY, 1978, p. 524.

113. L. Hsu and E. Y. Wang, in *13th IEEE Photovoltaic Spec. Conf.* New York, NY, 1978, p. 536.

113a. O. P. Agnihotri, A. Raza, and B. K. Gupta, unpublished.

114. L. L. Kazmerski and Peter Sheldon, in *13th IEEE Photovoltaic Spec. Conf.* New York, NY, 1978, p. 541.

115. T. Nagatomo and O. Omoto, *Jap. J. Appl. Phys.*, **15**, 199 (1976).

116. S. Franz, G. Kent, and R. L. Anderson, *J. Electron. Mat.*, **6**, 107 (1977).

117. W. G. Thomson, S. L. Franz, R. L. Anderson, and O. H. Winn, *IEEE Trans. Elect. Devices*, **ED-24**, 463 (1977).

118. T. R. Nash and R. L. Anderson, *IEEE Trans. Elect. Devices*, **ED-24**, 468 (1977).

119. A. K. Ghosh, C. Fishman and T. Feng, *J. Electrochem. Soc.*, **124**, 317C (1977).

120. R. H. Bube, in *Dept. of Energy Advanced Materials Res. and Devel. Branch Review Meetings*, Washington, D.C., April, 1978.

121. K. S. Sree Harsha, K. J. Bachmann, P. H. Schmidt, E. G. Spencer, and F. A. Thiel, *Appl. Phys. Lett.*, **30**, 645 (1977).

122. L. L. Kazmerski, in *Dept. of Energy Advanced Materials res and Devel. Branch Review Meetings*, Washington, D.C., April, 1978.

123. G. V. Samsonov, *The Oxide Handbook* (IFI/Plenum Data Company, New York, NY, 1973), p. 28.

124. R. Singh and J. Shewchun, *J. Appl. Phys.*, **49**, 4588 (1978).

125. K. Kaziyama and Y. Furukawa, *Jap. J. Appl. Phys.*, **6**, 905 (1967).

126. T. Nishino and Y. Hamakawa, *Jap. J. Appl. Phys.*, **9**, 1085 (1970).

127. T. Nishino and Y. Hamakawa, *Proc. Intl. Conf. on Heterojunctions*, Budapest, Hungary 1970 (Akademiai Kiado, Budapest, Hungary, 1971), Vol. 2, p. 409.

128. H. Matsunami, K. Co., H. Ito, and T. Tanaka, *Jap. J. Appl. Phys.*, **14**, 915 (1975).

129. H. Kato, A Yoshida, and T. Arizumi, *Jap. J. Appl. Phys.*, **15**, 1819 (1976).

130. R. L. Anderson, "Glass-Si Heterojunction Solar Cells," NSF Grant No. AER 74–17631, Final Report, August 31, 1975.

131. J. L. Vossen and E. S. Poloniak, *Thin Solid Films*, **13**, 281 (1972).

132. J. L. Vossen, U. S. Patent 3, 749, 658 (1973).

133. T. Mizrah, Sc. D. Thesis (MIT, Cambridge, MA, 1976).

134. R. Swalin, *Thermodynamics of Solids*, 2nd ed. (Wiley, New York, NY, 1972).

135. A. K. Ghosh C. Fishman, and T. Feng, *J. Appl. Phys.*, **49**, 3490 (1978).

136. T. Feng, C. Fishman, and A. K. Ghosh, in *13th IEEE Photovoltaic Spec. Conf.* Washington, D.C., 1978, p. 519.

137. L. Hsu and E. Y. Wang, in *Photovoltaic Solar Energy Conf.*, A. Strub, Ed. (D. Reidel, Dordrecht, Holland, 1977), p. 1100.

138. T. Nagatomo and D. Omoto, *Jap. J. Appl. Phys.*, **15**, 199 (1976).

139. D. E. Burk, J. B. DuBow, and J. R. Sites, in *Proc. 12th IEEE Photovoltaic Spec. Conf.*, New York, NY, 1976, p. 971.

140. E. Fabre and Y. Bandet, in *Photovoltaic Solar Energy Conf.*, A. Strub, Ed. (D. Reidel, Dordrecht, Holland, 1977), p. 178.

141. D. C. Carlson, *IEEE Trans. Elect. Devices* **ED-24**, 449 (1977).

142. M. Perotin, L. Szepessey, J. C. Manifacier, P. Parto, J. P. Fillard, and M. Savelli, *Solar Electricity* (Maison des Congres, Tononse, France, 1976), p. 481.

143. H. Kato, J. Fujimote, T. Kanada, A. Yoshida and T. Arizumi, *Phys. Stat. Sol.*, **a32**, 255 (1975).

144. T. Matsushita, A. Suzuki, M. Okuda, and T. T. Nang, *Jap. J. Appl. Phys.*, **16**, 2057 (1977).

145. T. T. Nang, M. Okuda, T. Matsushita, S. Yokota, and A. Suzuki, *Jap. J. Appl. Phys.*, **15**, 383 (1976).

146. M. Okuda. T. T. Nang, T. Matsushita, and S. Yokota, *Jap. J. Appl. Phys.*, **14**, 1597 (1975).

147. T. Matsushita, T. T. Nang, M. Okuda, A. Suzuki, and S. Yokota, *Jap. J. Appl. Phys.*, **15**, 901 (1976).

148. G. Cheek, D. Ellsworth, A. Genis, and J. Dubow, *Proc. 20th Elect. Mat. Conf.* (IEEE, New York, NY, 1978).

149. R. N. Legge, Ph. D. Thesis (Wayne State University, Detroit, MI, 1977).

150. D. L. Feucht, *J. Vac. Sci. Technol.*, **14**, 57 (1977).

151. C. W. Wilmsen and R. W. Kee, *J. Vac. Sci. Technol.*, **14**, 953 (1977).

152. W. G. Thomson and R. L. Anderson, *Sol. State Electro.* **21**, 603 (1978).

153. R. Singh and J. Shewchun, *Appl. Phys. Lett.*, **33**, 601 (1978).

154. W. G. Haines and R. H. Bube, *J. Appl. Phys.* **49**, 304 (1977).

155. E. Bucher, *Appl. Phys.* **17**, 1 (1978).

156. J. DuBow, G. Cheek, A. P. Genis, C. Wilmsen, J. F. Wager, and J. Shewchun, *Proc. IEEE 13th Photovoltaic Spec. Conf.* New York, NY, 1978, p. 767.

157. R. Siegel and J. R. Howell, *Thermal Radiation Heat Transfer* (McGraw Hill, New York, NY, 1972).

158. J. T. Bevans and D. K. Edwards, private communication.

159. R. V. Dunkle, in *Surface Effects on Spacecraft Materials*, F. J. Clauss, Ed. (Wiley, New York, NY, 1960), p. 123.

160. D. M. Gates, C. C. Shaw, and D. Beaumont, *J. Opt. Soc. Am.*, **48**, 88 (1958).

161. H. E. Bennett and W. E. Koehler, *J. Opt. Soc. Am.*, **50**, 1 (1960).

162. J. Strong, *Procedures in Experimental Physics* (Prentice Hall, Englewood Cliffs, NJ, 1938), p. 376.

163. L. F. Drummeter and G. Hass, in *Physics of Thin Films* Vol. 2, A. G. Hass and R. F. Thun, Eds. (Academic, New York, NY, 1964), p. 305.

164. D. K. Edwards, J. T. Gier, K. E. Nelson, and R. D. Roddick, *J. Opt. Soc. Am.*, **51**, 1279 (1961).

165. L. Harris and P. Fowler, *J. Opt. Soc. Am.*, **51**, 164 (1961).

166. J. T. Gier, R. V. Dunkle, and J. T. Bevans, *J. Opt. Soc. Am.*, **44**, 558 (1954).

167. J. T. Gier, R. H. Mansfield, R. V. Dunkle, N. W. Snyder, and L. Possner, Thermal Radiation Project Report No. 5, Contract No. NR-014-062, Institute of Eng. Research, Univ. of California, Berkeley, CA (1948).

168. R. V. Dunkle, in *1st Symp. Surface Effects on Spacecraft materials*, sponsored by the U.S. Air Force Research and Development Command, and Lockheed Missiles and Space Division, Pala Alto, CA, 1959, (Wiley New York, NY, 1959).

169. R. V. Dunkle, D. K. Edwards, J. T. Gier, and J. T. Bevans *Solar Energy*, **4**, 27 (1960).

170. H. J. McNicholas *J. Res. Natl. Bur. Std.* (U.S.) **1**, 29 (1928).

171. E. Karrer Sci. *Papers Natl. Bur. Std.* (U.S.) **17**, 203 (1922).

172. C. P. Tingwaldt, *Optik*, **9**, 323 (1952).

173. A. S. Toporets, *Opt. Spectry.* **7**, 471 (1959).

174. C. P. Tingwaldt, in *Trans. Conf. Use of Solar Energy* Tucson, AZ, October 31, 1955, E. F. Carpenter, Ed. (univ. of Arizona Press, Tucson, AZ, 1955), Vol. 2, Part 1, p. 57.

175. R. V. Dunkle, D. K. Edwards, J. T. Gier, K. E. Nelson, and R. D. Roddick, in *Prog. Inst. Res. Thermodynamic* and *Transport Prop.* (American Soc. of Mechan. Eng., 1962), p. 541.

176. J. T. Gier, R. V. Dunkle, and J. T. Bevans, Final Progress Report Snow Characteristics Project, Institute of Engineering Research, Univ. of California Berkeley, CA (1955).

177. S. T. Dunn, J. C. Richmond, and J. A. Wiebelt, *J. Opt. Soc. Am.*, **55**, 604 (1965).

178. S. T. Dunn, J. C. Richmond, and J. A. Wiebelt, *J. Spacecraft Rockets*, **3**, 961 (1966).

179. R. T. Neher and D. K. Edwards, *Appl. Optics*, **4**, 755 (1965).

180. D. K. Edwards, *Appl. Optics*, **5**, 175 (1966).

181. B. O. Seraphin, Chemical Vapor Deposition Research for Fabrication of Solar Energy Converter. Techn. Rept. NSF/RANN Grant SE/GI-36731 (1974).

182. R. Willey, Willey Corporation, Melbourne, Australia.

183. C. P. Butler and R. J. Jenkins, "Measurement of Thermal Radiation Properties of Solids," J. C. Richmond, Ed. NASA SP-31 (1963), p. 39.

184. E. R. Streed, L. A. McKellar, R. J. Rolling, and C. A. Smith, "Measurement of Thermal Radiation Properties of Solids," J. C. Richmond Ed., NASA SP-31 (1963) p. 237.

185. J. T. Neu. "Design, Fabrication and Performance of an Ellipsoidal Spectrophotometer," NASA CR-73193 (1968).

186. R. E. Gaumer, G. F. Hohnstreiter, and G. F. Vanderschmidt, "Measurement of Thermal Radiation Properties of Solids," J. C. Richard, Ed., NASA SP-31 (1963), p. 117.

187. K. E. Nelson, E. E. Luedke and J. T. Bevans, *J. Spacecraft Rockets*, **3**, 758 (1966).

188. W. B. Fussell, J. J. Triolo, and J. H. Henniger, "Measurement of Thermal Properties of Solids," J. C. Richmond, Ed., NASA SP-31 (1963).

189. R. B. Pettit, 75.0079, Sandia Laboratories, (1975).

190. Y. S. Touloukian, D. P. Dewitt, and R. S. Hernicz, Eds., *Thermal Radiative Properties of Coatings* (Plenum Press, New York, NY, 1972), p. 533.

191. Y. S. Touloukian, D. P. Dewitt, and R. S. Hernicz, Eds., *Thermal Radiative Properties of Coatings* (Plenum Press, New York, NY, 1972), p. 121.

192. R. Sadler, L. Hemmerdinger, and I. Rando, "Measurement of Thermal Properties of Solids," J. C. Richmond, Ed., NASA SP-31 (1963), p. 217.

193. G. A. Zerlaut, "Measurement of Thermal Properties of Solids," J. C. Richmond, Ed., NASA SP-31 (1963), p. 275.

194. R. E. Gaumer and J. V. Stewart, "Measurement of Thermal Properties of Solids," J. C. Richmond, Ed., NASA SP-31 (1963), p. 127.

195. H. Willrath and R. B. Gammon, *Solar Energy*, **21**, 193 (1978).

196. D. B. McKenney, W. T. Beauchamp NSF/RANN/SE/GI-41895/PR/74/4 NTIS, U.S. Department Commerce, Washington D.C., (1974).

197. G. E. McDonald, Techn. Rept. NASA TMX-73697 (1977).

198. R. R. Willey, *Appl. Spectry.*, **30**, 593 (1976).

199. R. R. Willey, *AES Coating for Solar Collector Symp.*, Atlanta, GA, 1976, p. 9.

199a. R. B. Pettit, "Trans. ASME J. Eng. for Power," Paper No. 77-WA/SOL-I (1978).

200. B. O. Seraphin, in *Topics in Applied Physics*, Vol. 31, B. O. Seraphin, Ed. (Springer-Verlag, Berlin, Germany, 1979).

201. B. O. Seraphin in *Physics of Thin Films*, Vol. 10, G. Hass and M. H. Francombe, Eds. (Academic, New York, NY, 1979), p. 1.

202. A. F. Turner, private communication.

203. H. Ehrenreich and B. O. Seraphin, *Rept. Symp. Fundamental Opt. prop. Solids Relevant to Solar Energy Conversion*, Tucson, AZ, 1975.

204. Y. S. Touloukian, D. P. Dewitt, and R. S. Hernicz, *Thermophysical Properties of Matter*, Vol. 9, IFI/(Plenum Data Company, New York, 1972).

205. E. Kauer, U.S. Patent 2, 228, 625 (1966).

206. A. B. Meinel and M. P. Meinel, *Applied Solar Energy*, Addison-Wesley, Reading, MA, 1977).

207. H. Tabor, "Selective Surfaces for Solar Collector, Ch. IV, Low Temperature Engineering Applications of Solar Energy, "American Society of Heating, Refrigeration, and air-conditioning Engineering (ASHRAE) (1967).

208. H. Tabor, J. Harris, H. Wenberger, and B. Doron, *Further Studies on Selective Black Coatings, UN Conf. New Sources of Energy*, Rome, Italy, 1961.

209. H. Tabor, U.S. Patent 2,917,817 (1955).

210. G. E. McDonald and H. B. Curtis, Techn. Rept. No. NASA-TMX-73498 (1976).

211. H. Y. B. Mar, R. J. H. Lin, P. B. Zimmer, R. E. Peterson, and J. S. Gross, Techn. Rept. No. PB 252–383, Honeywell Corporation (1976).

212. J. T. Borzoni, *AES Coatings for Solar Collector Symp.*, Atlanta GA, 1976, p. 89.

213. R. R. Sowell and R. B. Pettit, Techn. Rept. No. SAND 78–0554, Sandia Laboratories (1978).

214. M. Ramakrishna Rao, J. Balachandra, and K. I. Vasu. *Proc. ISES*, New Delhi, India, January 1978, F. deWinter and M. Cox, Eds. (Pergamon Press, New York, NY, 1978), p. 875.

215. P. K. Gogna, D. K. Pandya, and K. L. Chopra, *Proc. ISES*, New Delhi, India, January 1978, F. deWinter and M. Cox, Eds., (Pergamon Press, New York, NY, 1978), p. 842.

216. P. K. Gogna and K. L. Chopra, *Thin Solid Films*, **57**, 299 (1979).

217. G. E. McDonald and H. B. Curtis, Techn. Rep. TMX-71731, NASA (1977).

218. G. E. McDonald, Techn. Rept. NASA-71596 (1976).

219. G. E. McDonald, Techn. Rep. NASA-73799 (1977).

220. H. Y. B. Mar, R. E. Peterson, and P. B. Zimmer, *Thin Solid Films*, **39**, 95 (1976).

221. R. J. H. Lin and P. B. Zimmer, Techn. Rept. COO–2930–12, Honeywell Carporation (1977).

222. Products of Harshaw Chemical Company.

223. B. K. Duramir, "Black Chromium Process," E. I. Dupont De Nemours and Co., Inc., Industrial Chemical Department, Wilmington, DE 19898.

224. Metal Finishing Guide Book Directory, Metals & Plastics Publications, Inc. Westwood N. J. (1972), p. 628; G. E. McDonald, B. Buzek, and H. B. Curtis, Techn. Rept. TMX-73461 (1976).

225. H. Tabor, *Trans. Conf. Use of Solar Energy*, Tucson, AZ, October 31, 1955, E. F. Carpenter, Ed. (Univ. of Arizona press, Tucson, AZ, 1955), Vol. 2, p. 24.

226. G. E. McDonald, B. Buzek, and H. Curtis, Techn. Rept. NASA, Lewis is Res. Center. TMX-73461 (1976).

227. R. B. Pettit and R. R. Sowell, *J. Vac. Sci. Technol.*, **13**, 596 (1976).

228. Harshaw Chromonyx Black Chromium Process, Harshaw Chemical Co., Div. of Kewanee Oil Co., Cleveland OH.

229. S. W. Hogg and G. B. Smith, *J. Phys.* **D, 10**, 1863 (1977).

230. G. L. Harding, *Thin Solid Films*, **38**, 109 (1976).

231. A. Ignatiev, P. O'Neill, and G. Zajac, *Solar Energy Mat.*, **1**, 69 (1979).

232. H. Tabor, *Bull. Res. Council Israel* **5A**, 119 (1956).

232a. M. C. Keeling, R. K. Asher, and R. W. Gurtler, *AES Symp. Coatings for Solar Collector, Atlanta, GA* (1976).

233. D. J. Close, CSIRO Eng. Soc. Report ED7 (1962), Melbourne, Australia.

233a. G. E. Mc Donald R. Lauver, and H. B. Curtis, *Sharing the Sun* V. 6, *Photovoltaic and Material*, Winnipeg, Canada, 1976, K. W. Boer, Ed., p. 162.

234. E. Salem and F. Daniels *Solar Energy*, **3**, 19 (1959).

235. H. C. Hottel and T. A. Unger, *Solar Energy*, **3**, 10 (1959).

236. E. A. Christie, *Proc. ISES*, Melbourne, Australia, 1970, Paper No. 7181.

237. J. Hajdu and F. Brindisi Jr., *AES Symp. Coatings for Solar Collectors, Atlanda*, GA., (1976) p. 29.

238. C. V. Bishop & R. Dargis, *AES Symp. Coatings for Solar Collectors*, Atlanta, GA. (1976) p. 109.

239. R. S. Lindstrom, R. L. Merriam, E. H. Newton, and G. Cypher, *AES Symp. Coatings for Solar Collectors*, Atlanta, GA, (1976) p. 113.

240. A. F. Reid and A. F. Wilson, "The selective surface films on copper absorber plates," Division of Mineral Chemistry, CSIRO, Melbourne Australia, private communication.

241. P. M. Driver & P. G. McCormik, *Proc. ISES*, New Delhi, India January, 1978, F. de Winter and M. Cox, Eds. (Pergamon Press, New York, NY, 1978) p. 881.

242. D. M. Mattox and R. R. Sowell, *J. Vac. Sci. Technol.*, **11**, 793 (1974).

243. D. K. Edwards and R. Roddick, Spectral and Directional Thermal Radiation Characteristics of Selective surfaces for solar collectors" J. Solar Eng. Sci. & Engg. 6 (1962).

244. J. A. Duffie, *Solar Energy*, **6**, 114 (1962).

245. G. Hass, H. H. Schraeder, and A. F. Turner, *J. Opt. Soc. Am.*, **46**, 31 (1956).

246. F. Trombe, M. Foex, and M. le Phat Vin, *U N Conf. on New Sources of Energy*, Rome, Italy, 1961, p. 421.

247. Presto Black Conversion finish for Ferrous Metals, Birchwood Casey Div. of Fuller Laboratories Inc., Edn Prairie, Minnesota.

248. Enthone Inc., A. Subsidiary of American Smelting and Refining Co., Bridgeview, Il.

249. Dow Corning Corp., Midland, Mi 48640.

250. Exxon Chemical Co., Houston, TX.

251. M. Vander Leij, *Proc. ISES*, New Delhi India, January, 1978, F. deWinter and M. Cox, Eds. (Pergamon Press, New York, NY, 1978), p. 837.

252. M. Vander Leij, *J. Electrochem. Soc.*, **125**, 1361 (1978).

253. R. D. Srivastava and S. Kumar, *Plating*, **60**, 487 (1973).

254. J. A. Thronton, *AES Coatings for Solar collector symp.*, Atlanta, GA, 1976, p. 63.

255. G. E. McDonald and H. B. Curtis, Techn. Rept. NASA-TMX-71817 (1975).

256. M. A. Encheva, *J. Appl. Chem. USSR*, **45**, 318 (1972).

257. L. F. G. Williams, *Plating*, **59**, 931 (1972).

258. B. Karlsson and C. G. Ribbing, *Proc. Soc. Photo-Optical Eng.* **161** (4), 161/09 (1978).

259. B. Karlsson and C. G. Ribbing, Techn. Rept. NSF Contract 3879–1 (1977).

260. G. B. Smith, *Metal Austral.*, 204, (September 1977).

261. G. Granziera, *Metal Austral.*, 211 (September 1977).

262. C. E. Batcheller, *Metal Finn.* **42**, 466 (1944).

263. T. E. Evans, A. C. Hart, H. James, and V. A. Smith, *Trans. Inst. Metal Finn.*, **50**, 77 (1972).

264. T. E. Evans, A. C. Hart, A. N. Skedgell, *Trans. Inst. Metal Finn*, **51**, 108 (1973).

265. G. E. Carver, H. S. Gurev, and B. O. Seraphin, *J. Electrochem. Soc.*, **125**, 1138 (1978).

266. J. H. Powers, A. G. Craig, Jr., and W. King, Joint Conference American Section, *Proc. ISES and SES Canada*, V. 6, Winnipeg, Canada, August 1976, K. W. Boer, Ed., p. 166.

267. W. C. Cochran and J. H. Powers. Techn. Serv. Rept. No. 216, Alcoa Center (1976).

268. D. G. Altenpohl, *Use of Boehmite Films for Corrosion Protection of Aluminium, Corrosion*, 143 (1962).

269. H. P. Godard, *The Corrosion of Light Metals* (Wiley, New York, NY, 1967), p. 45.

270. K. Kimoto and I. Nishida, *Jap. J. Appl. Phys.*, **6**, 1047 (1967).

271. D. M. Mattox and G. J. Kominiak *J. Vac. Sci. Technol.*, **12**, 182 (1975).

272. D. M. Mattox, "Sputter Deposition and Ion Plating Technology," Monograph, Thin Film Division of American Vacuum Society, New York, NY, (1973).

273. H. S. Gurev and B. O. Seraphin, *Proc. 5th Intl. Conf.* CVD, Buckinghamshire, England, 1975 (J. M. Blocher, Jr., H. E. Hinterman, and L. H. Hall, Eds., (The Electrochemical Soc. Inc.), p. 667.

274. R. E. Hahn and B. O. Seraphin, *J. Vac. Sci. Technol.*, **12**, 905 (1975).

275. A. E. B. Presland, G. L. Price, and D. L. Trimm, *Surf. Sci.* **29**, 424, 435 (1972).

276. B. O. Seraphin and V. A. Wells, *Intl. Conf. The Sun in Mankind Service*, Paris, France, 1975.

277. A. Donnadieu and B. O. Seraphin, *J. Opt. Soc. Am.*, **68**, 292 (1978).

278. D. A. Williams, T. A. Lappin, and J. A. Duffie, *J. Eng. Power* **85A**, 213 (1963).

279. T. J. McMahon and S. N. Jasperson, *Appl. Optics*, **13**, 2750 (1974).

280. R. Marchini and R. Gandy, *J. Appl. Phys.*, **49**, 390 (1978).

281. O. H. Olson, *Appl. Optics*, **2**, 109 (1963).

282. B. K. Gupta, R. Thangaraj, and O. P. Agnihotri, *Solar Energy Mat.*, **1**, 471 (1979).

283. R. Thangaraj, B. K. Gupta, A. K. Sharma, O. P. Agnihotri, and S. S. Mathur, Techn. Rept. TRI CES/S-10, Indian Institute of Technology, New Delhi (1978).

284. G. L. Harding, *J. Vac. Sci. Technol.* **15**, 65 (1978).

285. G. L. Harding, *J. Vac. Sci. Technol.*, **13**, 1070 (1976).

286. G. L. Harding, *J. Vac. Sci. Technol.*, **13**, 1073 (1976).

287. G. L. Harding, *J. Vac. Sci. Technol.*, **14**, 1313 (1977).

288. L. Geoffrey, G. L. Harding, and I. T. Ritchie, *Proc. ISES* New Eelhi, India, January 1978, F. de Winter and M. Cox, Eds. (Pergamon Press, New York, NY, 1978), p. 845.

289. G. V. Samsonov, *Handbook of High Temperature Materials* (Plenum Press, New York, NY, 1964), p. 94.

290. R. Blickernsderfer, R. L. Lincoln, and D. K. Deardorff, "Reflectiance and Emittance of Spectrally Selective Titanium and Zirconium Nitrides," Techn. Rept. 8167, Bureau of Mines, Washington, D.C. (1976).

291. A. Goldsmith, T. Waterman, and H. Hirschoorn, *Handbook of Thermophysical Properties of Solid Materials*, Vol. 4 (Macmillan, New York, NY, (1961).

292. R. R. Sowell and D. M. Mattox, in *SES Coatings for Solar Collector Symp.*, Atlanta, GA, 1976, p. 22.

293. J. C. C. Fan and S. A. Spura, *Appl. Phys. Lett.*, **30**, 511 (1977).

294. M. Telkes, U.S. Patent 4,011,190 (1977).

295. B. K. Gupta, F. K. Tiwari, O. P. Agnihotri, and S. S. Mathur. *Intal. J. Energy Res.* **3**, 371 (1979).

296. Ferro Corporation, Cleveland, OH.

297. K. C. Park, *Appl. Optics*, **3**, 877 (1964).

298. R. N. Schmidt, K. C. Park, and J. E. Janssen, "High Temperature Solar Absorber Coatings," Part II, ML-TDR-64-250 September, Honeywell Corporation (1964).

299. R. N. Schmidt, K. C. Park, R. H. Torborg, and J. E. Janssen, "High Temperature Solar Absorbing Coatings," Part I, AST-TDR-63–579, Honeywell Carporation (1963).

300. A. Sh. Sharafi and A. G. Mukminova, *Appl. Solar Energy* **11.2**, 99 (1975).

301. A. Sh. Sharafi and A. G. Muminova, Nauchynitrudy Tashgu Fizika No. 413 (1971).

302. C. N. Watson-Munro and C. M. Horwitz, in *Solar Energy*, H. Messil and S. T. Butter, Eds. (Pergamon Press, New York, NY, 1975), Chap. 8, p. 293.

303. R. N. Schmidt and K. C. Park, *Appl. Optics*, **4**, 917 (1965).

304. H. Y. B. Mar, "Optical coating for Flat Plate Collector," COO/2625/75/1 ERDA (1975).

305. A. B. Meinel, D. B. McKenney, and W. T. Beauchamp, NSF/RANN/SE/GI-41895/PR/74/4, Washington, D.C. (1974).

306. V. V. Li, A. Faiziev, V. Kh. Gaziev, and U. S. Trukhov, *Appl. Solar Energy* **13.6**, 40 (1977).

307. R. E. Peterson and J. W. Ramsey, *J. Vac. Sci. Technol.*, **12**, 174 (1975).

308. F. Trombe and M. Foex, *CNRS Appl. Therm. Energ. Sol. Symp.*, Montlouis, France, 1958, p. 621.

309. F. Trombe, M. Foex, and M. le Phat Vin, *UN Conf. on New Sources of Energy*, Rome, Italy, 1961.

310. K. G. T. Holland, *Solar Energy*, **7**, 117 (1963).

311. D. J. Close, *Solar Energy*, **7**, 117 (1963).

312. J. J. Cuomo, J. F. Ziegler, and J. M. Woodale, *Appl. Phys. Lett*, **26**, 557 (1975).

313. G. D. Pettit, J. J. Cuomo, T. H. DiStefano, and J. M. Woodall, *IBM Res. Rev.* **22**, 372 (1978).

314. T. F. Irvine Jr., J. P. Hernett, and E. R.G. Eckert, *Solar Energy*, **2**, 19 (1958).

315. R. B. Pettit and R. R. Sowell, "Solar Absorptance and Emittance Properties of Several Solar Coatings," Report SAND-75-5066 Sandia Labs. Albuquerque, NM (1975).

316. G. K. Wehner and D. J. Hajicek, *J. Appl. Phys.* **42**, 1145 (1971).

317. W. R. Hudsen, *J. Vac. Sci. Technol.*, **14**, 286 (1977).

318. D. P. Grimmer, K. C. Herr, and W. J. McCreary, *J. Vac. Sci. Technol.*, **15**, 59 (1978).

319. J. J. Cuomo, J. M. Woodall, and T. H. DiStefano, *AES Coating for Solar Collector Symp.* Atlanta, GA, 1976, p. 133.

320. T. Santala and R. Sabol, "Development of High Efficiency Collector Plates," Texas Instruments Report, COO/2600-73/3, Attleboro, MA, (1976).

321. L. Harris and J. K. Beasley, *J. Opt. Soc. Am.* **42**, 134 (1952).

322. L. Harris, R. J. McGinnies, and B. M. Siegel, *J. Opt. Soc. Am.*, **38**, 582 (1948).

323. D. R. McKenzie, *J. Opt. Soc. Am.*, **66**, 249 (1976).

324. D. R. McKenzie, *Gold Bull.*, **2**, 49 (1978).

325. D. R. McKenzie, *Appl. Phys. Lett.*, **34**, 25 (1978).

326. D. M. Mattox, *J. Vac. Sci. Technol.*, **13**, 127 (1976).

327. P. B. Johnson and R. W. Christy, *Phys. Rev.* **B9**, 5056 (1974).

328. G. E. McDonald. Techn. Rept. NASA-TMX-3136 (1974).

329. A. J. Sievers, *Materials Research Council Conf.*, 1973, La Jolla, CA, Sponsored by ARPA, Washington, Vol. 2, p. 111.

330. W. Hampe, *Z. Phys.* **152**, 470, 476 (1958).

331. G. Rasigni and P. Rouard. *J. Opt. Soc. Am.*, **53**, 604 (1963).

332. S. Yoshida, T. Yamaguchi, and A. Kinbara, *J. Opt. Soc. Am.*, **61**, 62 (1971).

333. S. Yoshida, T. Yamaguchi, and A. Kinbara, *J. Opt. Soc. Am.*, **61**, 463 (1971).

334. S. Yoshida, T. Yamaguchi, and A. Kinbara, *J. Opt. Soc. Am.*, **62**, 634 (1972).

335. S. Yoshida, T. Yamaguchi, and A. Kinbara, *J. Opt. Soc. Am.*, **62**, 1415 (1972).

336. R. H. Doremus, *J. Chem. Phys.*, **40**, 2389 (1964).

337. R. H. Doremus, *J. Chem. Phys.*, **42**, 414 (1965).

338. R. H. Doremus, *J. Appl. Phys.*, **37**, 2775 (1966).

339. N. C. Miller, B. Hardiman, and G. A. Shim, *J. Appl. Phys.* **41**, 1850 (1970).

340. R. W. Cohen, G. D. Cody., M. D. Coutts, and B. Abeles, *Phys. Rev.* **B8**, 3689 (1973).

341. J. I. Gittleman, B. Abeles, P. Zanzucchi, and Y. Arie, *Thin Solid Films*, **45**, 9 (1977).

342. J. I. Gittleman, *Appl. Phys. Lett.*, **28**, 370 (1976).

343. J. C. C. Fan and P. M. Zavracky, *Appl. Phys. Lett.*, **29**, 478 (1977).

344. B. Abeles and J. I. Gittleman, *Appl. Optics*, **15**, 2328 (1976).

345. D. R. McKenzie and R. C. McPhedran, in *Electrical Transport and Optical Properties of Inhomogeneous Media, AIP Conf. Proc. No. 40*, J. C. Garland and D. W. Tanner, Eds. (AIP, New York, NY, 1978), p. 283.

346. P. H. Lissberger and R. G. Nelson, *Thin Solid Films*, **21**, 159 (1974).

347. J. C. M. Garnett, *Phil. Trans. Royal Soc. London*, **203**, 385 (1904).

348. W. Lamb, D. M. Wood, and N. W. Ashcroft, in *Electrical Transport and Optical Properties of Inhomogeneous Media, AIP Conf. Proc. No. 40*, J. C. Garland and D. W. Tanner, Eds. (AIP, New York, NY, 1978), p. 240.

349. J. C. M. Garnett, *Phil. Trans. Royal Soc. London*, **205**, 237 (1906).

350. L. Genzel and T. P. Martin, *Surf. Sci.* **34**, 33 (1973).

351. D. A. G. Bruggeman, *Ann. Phys.* **24**, 636 (1935).

352. C. G. Granqvist. *J. Appl. Phys.*, **50**, 2961 (1979).

353. D. R. McKenzie and R. C. McPhedran, in *Electrical Transport and Optical Properties of Inhomogeneous Media, AIP Conf. Proc.*, J. C. Garland and D. W. Tanner, Eds. (AIP, New York, NY, 1978), p. 283.

354. P. H. Lissberger and R. G. Nelson, *Thin Solid Films*, **21**, 159 (1974).

355. P. M. Driver, R. W. Jones, C. L. Riddiford, and R. J. Simpson, Solar Energy, **19**, 301 (1977).

356. C. G. Granqvist and O. Hunderi, *J. Appl. Phys.*, **50**, 1058 (1979).

357. C. G. Granqvist, *Intl. Symp. on Solar Thermal Power Stations*, Cologne, Germany, April, 1978.

358. H. G. Craighead and R. A. Buhrman, *J. Vac. Sci. Technol.*, **15**, 269 (1978).

359. H. G. Craighead and R. A. Buhrman, *Appl. Phys. Lett.*, **31**, 423 (1977).

360. H. G. Craighead, R. Bartynski, R. A. Buhrman, L. Wojcik, and A. J. Sievers, *Solar Energy Mat.* **1**, (1979).

361. J. I. Gittleman, B. Abeles, P. Zanzucchi, and Y. Arie, *Thin Solid Films*, **45**, 9 (1977).

362. J. I. Gittleman, E. K. Sichel and Y. Arie, *Sol. Energy Mat.*, **1**, 93 (1979).

363. B. Abeles, *Granular Metal Films in Applied Solid State Science* Vol. 6, R. Wolf, Ed. (Academic, New York, 1976), p. 1.

364. L. D. Landau and E. M. Lifshitz, *Electrodynamics of Continuous Media* (Addison-Wesley, Reading, MA, 1960), pp. 20, 42.

365. V. I. Odelovskii, *J. Tech. Phys.*, **21**, 678 (1951).

366. R. Landauer, *J. Appl. Phys.*, **23**, 77a (1952).

367. D. Polder and J. H. VanSantan, *Physica*, **12**, 257 (1946).

368. O. Hunderi, *Phys. Rev.*, **B7**, 3419 (1973).

369. G. Burrafato, G. Giaquinta, N. A. Mancini, and A. Pennisi, in *Proc. 1st Course on Solar Energy Conversion*, 1975, Procida, Italy N. A. Mancini and I. F. Quercia, Eds.

370. G. Burrafato, *Heliotechnique and Development*, Vol. 1 M. A. Ketani and J. E. Sousou, Eds. (Development Analysis Association, Inc., 1977).

371. G. Giaquinta and N. A. Mancini, *Proc. 1st Izmir Cong. on Solar Energy*, Izmir Turkey August 1977.

372. G. Barrafato, G. Giaquinta, N. A. Mancini A. Pennisi, and S. O. Troia, *Pro. comp. Inter Meeting*, Thahran, Saudi Arabia, November 1976.

373. G. Giaquinta and N. A. Mancini, *Intl. Symp. Workshop on Solar Energy*, Cairo, Egypt, June 1978.

374. J. A. Thornton, *SEA Trans.*, **B2**, Sec. 3 1787 (1974).

375. V. B. Veinberg, *Optics in Equipment for the Utilization of Solar Energy* (State Publishing House of Defence Industry, Moscow, USSR, 1929), p. 144.

376. G. Francia, *U. N. Conf. on New Sources of Energy*, Rome, Italy 1961, Vol. 4, p. 572.

377. K. G. T. Hollands, *Solar Energy*, **9**, 159 (1955).

378. H. Buchberg, O. A. Lalude, and D. K. Edwards, *Solar Energy*, **13**, 193, (1971).

379. C. L. Gupta and H. P. Garg, *Solar Energy*, **11**, 25 (1967).

380. G. R. Cunnington and E. R. Streed, Techn. Rept. Lockheed Missiles and Space Co. Inc., Palo Alto, CA (1973).

381. J. D. Mackenzie, *Proc. Workshop on Solar Collectors for Heating and Cooling of Buildings*, New York, NY, November 1974, p. 250.

382. P. Call, Proc. *DOE/DST Thermal Power Systems Workshop on Selective Absorber Coatings* December 1977, Paper SERI/TP-31-061.

383. P. Call, "National Program Plan for Absorber Surfaces," R and D Deptt. of Energy SERI/TR-31-103 January (1979).

384. L. E. Flordal and R. Kivaisi, *Vacuum*, **27**, 3971 (1977).

Author Index

Subject Index

HETERICK MEMORIAL LIBRARY
621.47 A273s
onuu
Agnihotri, O. P./Solar selective surface
3 5111 00108 0195